THE
TRAIN
BOOK

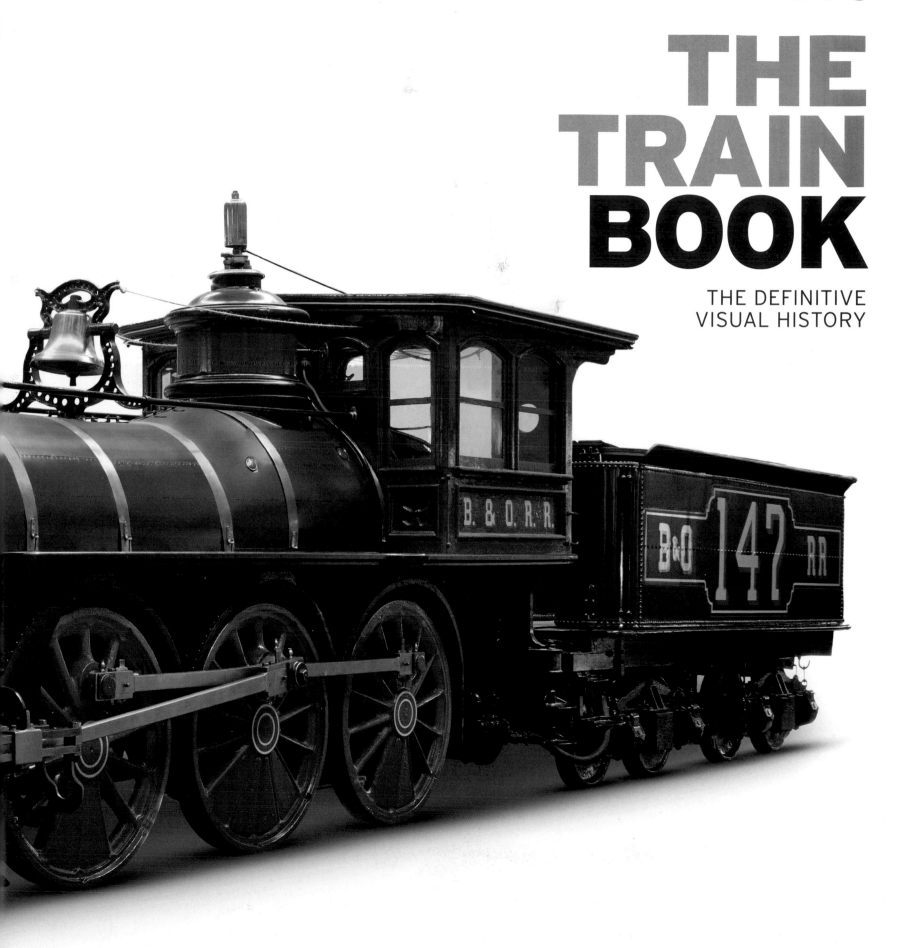

THE
TRAIN
BOOK

THE DEFINITIVE
VISUAL HISTORY

LONDON, NEW YORK, MELBOURNE, MUNICH, AND DELHI

DORLING KINDERSLEY

Senior Editors Sam Atkinson, Jemima Dunne, Kathryn Hennessy
Senior Art Editor Sharon Spencer
Project Art Editor Amy Child
Editors Suhel Ahmed, Rod Green, Alison Sturgeon, Miezen van Zyl
Editorial Assistance Alexandra Beeden
Design Assistance Alex Lloyd
Photographer Gary Ombler
Picture Research Nic Dean
DK Picture Library Claire Bowers, Claire Cordier, Romaine Werblow
Jacket Designers Amy Child, Mark Cavanagh
Jacket Editor Maud Whitney
Jacket Design Development Manager Sophia MTT
Producer, Pre-Production Nikoleta Parasaki
Producer Linda Dare
Managing Editor Esther Ripley
Managing Art Editor Karen Self
Publisher Sarah Larter
Art Director Phil Ormerod
Associate Publishing Director Liz Wheeler
Publishing Director Jonathan Metcalf

DK INDIA

Managing Editors Pakshalika Jayaprakash, Rohan Sinha
Managing Art Editors Arunesh Talapatra, Sudakshina Basu
Senior Editor Anita Kahar
Senior Art Editors Chhaya Sajwan, Mahua Sharma
Project Editor Antara Moitra
Project Art Editor Vaibhav Rastogi
Editor Vibha Malhotra
Art Editors Namita, Supriya Mahajan, Divya PR, Devan Das
Assistant Art Editors Roshni Kapur, Vansh Kohli, Riti Sodhi
Production Manager Pankaj Sharma
Pre-production Manager Balwant Singh
Senior DTP Designers Sachin Singh, Jagtar Singh
DTP Designers Nand Kishor Acharya, Bimlesh Tiwary
Picture Researcher Aditya Katyal
Picture Research Manager Taiyaba Khatoon

General Consultant Tony Streeter
Contributors Julian Holland, Keith Fender
Gary Boyd-Hope, Jonathan Randle Falconer, Peter Herring, Keith Langston,
Ashwani Lohani, Malcolm McKay, David Wilcock

First published in Great Britain in 2014 by
Dorling Kindersley Limited,
80 Strand, London WC2R 0RL

A Penguin Company

Copyright © 2014 Dorling Kindersley Limited

2 4 6 8 10 9 7 5 3 1
001 – 256473 – 10/14

A CIP catalogue record for this
book is available from the British Library.

ISBN: 978-1-4093-4796-5

Printed and bound in China by Leo Paper Products Ltd

Discover more at
www.dk.com

Contents

1804-1838: THE IRON HORSE

The invention of the steam-powered locomotive led to the development of the first passenger railway in Britain. This new mode of transport spread to other countries, with *Rocket* setting the benchmark for future locomotives.

1839-1869: BUILDING NATIONS

New tracks were laid across Europe, the US, and India. Meanwhile, engineers made further innovations to all aspects of rail travel, increasing its speed and efficiency. Mass city transit began with the London Underground.

1870–1894:
A WORLD OF STEAM

The rapid growth of the railway defined the power of human endeavour. Tracks negotiated every terrain and all kinds of obstacles, covering vast distances and making rail travel across continents possible. The glamour of rail travel was epitomized by grand stations and luxury services.

1895–1913: GOLDEN AGE

Electric-powered railways came into prominence in North America and Europe, while new innovations increased the efficiency of steam. Emulating London, Paris and New York introduced their own underground systems.

1914–1939: STEAM'S ZENITH

During World War I locomotives were key in the transport of soldiers and munitions. After hostilities ended, steam trains became faster and streamlined, and diesel trains were rolled out for the first time in the US and Europe.

1940–1959: WAR AND PEACE

The destruction of many European rail lines during World War II and the redrawing of national borders at the end of the conflict forced many governments to overhaul their rail systems. Technological advances saw diesel- and electric-power take over from steam.

1960–1979: BUILT FOR SPEED

The Japanese "bullet" train heralded a new age of high-speed rail travel, inspiring Western countries to innovation on their own railways. Increasing competition from road and air led to further modernization.

1980-1999: CHANGING TRACKS

New technology focused on developing high-speed networks throughout the world, but the period also saw the introduction of luxury trains. The Channel Tunnel opened, linking Britain to mainland Europe.

AFTER 2000: RAILWAY REVIVAL

The new millennium has seen China become a major proponent of rail travel, building tracks at an unprecedented rate and introducing new trains, including the ultrafast Maglev. On a global level, rail travel offered a more glamorous and luxurious alternative to the jetliner.

HOW RAILWAYS WORK: ENGINES AND TRACKS

This chapter offers an overview of basic rail technology, from how rails and locomotive wheels are designed, to signalling systems past and present. The engineering principles behind steam, diesel, and electric locomotives are explained.

The Railway Revolution

The click-clack of wheels on rails, the whiff of coal smoke and oil, a whistle in the distance, the feeling of anticipation and excitement at the start of a long journey ...

Railways capture our imagination. They speak to our soul. The elemental attractions of fire and steam, the fascination of technology, and the glamour of connecting faraway places have all helped cement the place of railways in human hearts. For more than 200 years, trains have fuelled ambitions and attracted ground-breaking engineers, inspiring them to create inventions that tapped into the human desire to move forward and open up a world of possibilities.

Most importantly, railways have contributed to modern history in prosaic, practical ways. Arguably, no single tool has influenced today's industrial world more. From the first stuttering experiments in Cornwall and Wales in the UK to the building of railways that opened up whole continents and helped create nations, as they did in North America and elsewhere, to their capacity to make modern warfare feasible – the invention of the locomotive has shaped the globe, for good and bad.

Before the railways, life moved at a different speed; most people travelled only short distances from where they lived – there were no cars, no planes, no modern roads. Until the arrival of trains there was no unified time and no compelling reason to introduce it. Towns and cities set their own time until the need for rigid timetables on the railways called for standardization. The new technology fuelled urbanization – growing conurbations were fed by railways, delivering people cheaply from ever farther afield. Rail networks moved commodities that previously could not be transported long distances – perishable fruit, newspapers, flowers, and fresh milk were delivered to the masses in a timely manner.

In these many ways, railways became essential to the creation of modern life, and achieved it with panache. Companies gave their locomotives and services evocative names; they came up with attractive colour schemes; and they worked hard on aesthetics to make their engines graceful, imposing, or dynamic, as well as functional. The drive to move ever forwards shaped the railways too. As new technologies developed, builders of new routes climbed higher, dug deeper, and went farther, taming the most inhospitable ground. The push to be ever faster, ever safer, and ever-more efficient drove that progress too.

Across the globe, railways put great effort into achieving higher speeds, into selling the luxury of their most exclusive trains, and into persuading people to use their services both for business and leisure. Modern marketing, public relations, the seaside holiday – in all these areas, the railway has been an instrument of change and a driving force. It is no wonder

that schoolboys have dreamt of becoming locomotive drivers, that authors as diverse as Leo Tolstoy, Émile Zola, Agatha Christie, and Sir Arthur Conan Doyle have bound railways into their dramas and mysteries, or that popular train-based songs like "Chattanooga Choo Choo" and "The Loco-Motion" have stood the test of time.

In the "Golden Age" of rail travel, newspapers and newsreels breathlessly reported the latest advances – as well as the gory details of smashes. New express engine designs were described in detail, drivers and designers became heroes, and there was fierce competition for headlines. Locomotives such as the huge "Big Boy" class in the US, or Britain's *Mallard* – which broke the speed record for steam in 1938 and still holds it – became famous the world over. Half a century after steam disappeared across large parts of the globe, it is still an emotive force – even among those who are too young to remember it in service. Dedicated enthusiasts chase the final survivors of the steam age in the most inaccessible places, or restore and preserve engines, coaches, and even entire lines.

It's not all been positive, and there's no denying the darker deeds that were made possible by railways. They offered an opportunity for mass transport that enabled huge armies to be moved and supplied across continents, as well as the deportation of millions of people to Hitler's extermination camps during World War II. War became global and more deadly, and it was inevitable that rail networks would themselves become targets and face huge destruction in modern conflicts.

Yet while railways entered increasingly difficult times – and after World War II a period came when they were often seen as bland, monotonous, and outdated – they always resisted becoming merely a thing of the past. In recent years there's been a renaissance as countries have directed energy into building new high-speed routes and reducing their reliance on the motor car.

Today, long container trains are still a vital component in freight delivery, rumbling across continents as the pioneering trains did more than 200 years ago. Passengers speed across borders without having to leave their seats, and the idea of moving people quickly over long distances in comfort is once again in vogue. Technology forges ahead, the glamour has returned, and, for many, railways are once again perceived as the civilized way to travel. Two centuries on from the pioneering moves of the first "iron horses", rail's exciting journey continues.

TONY STREETER
GENERAL CONSULTANT

1804–1838
THE IRON HORSE

THE IRON HORSE

In South Wales in February 1804, a new machine won a bet for its owner. The Pen-y-darren steam locomotive had just hauled wagons loaded with iron and people for nearly 10 miles (16 km). Richard Trevithick's machine was slow and cumbersome, but the achievement would soon change the world – the benefits of steam were felt quickly, and innovation was rapid.

In 1808 Trevithick's *Catch Me Who Can* pulled people around a short piece of circular track in London, but it was not until 1825 that passenger railways really began to take off. In September that year the world's first passenger line to be paid for by public subscription was opened between Stockton and Darlington in northeast England. Its first locomotive was *Locomotion No. 1*, created by the father-and-son team of George and Robert Stephenson. The new technology rapidly went international; in France, engineer Marc Seguin built his own locomotive in 1828, and in 1829 the British-built *Stourbridge Lion* brought the steam age to the US, on the Delaware and Hudson line.

Britain's famous Rainhill Trials were held in 1829 to decide which locomotives would be built for the world's first inter-city line, the Liverpool & Manchester Railway (1830). Triumph went to the Stephensons, with their *Rocket*. Many of the engine's innovations were so successful that *Rocket* set the basic layout for future locomotives right until the end of the steam era. Engineers on both sides of the Atlantic began to adapt designs for their own terrain. In 1830 the US's home-built *Tom Thumb* made its debut on the Baltimore & Ohio Railroad.

△ **Stephensons' engines**
The *Rocket* was far from the Stephensons' only locomotive success; *North Star* ran on the Great Western Railway between 1838 and 1871.

"The introduction of so **powerful an agent** as steam to a carriage on wheels will make **a great change** in the **situation of man**"

THOMAS JEFFERSON, US PRESIDENT

◁ **Rainhill's skew arch bridge** was opened in 1830, one year after the famous Rainhill Trials

Key Events

▷ **1804** A locomotive at Pen-y-darren Colliery, South Wales, launches the steam age. An earlier design by Trevithick had apparently been built, but little information survives.

▷ **1808** Trevithick's *Catch Me Who Can* is demonstrated in London.

▷ **1813-14** William Hedley builds *Puffing Billy* to run at Wylam colliery in northeast England.

▷ **1814** George Stephenson constructs his first locomotive, for Killingworth Colliery near Newcastle-upon-Tyne. It gains the name *Blücher*.

▷ **1825** The opening of Britain's Stockton & Darlington Railway, the world's first passenger railway paid for by public subscription.

▷ **1828** Marc Seguin builds France's first locomotive.

▷ **1829** *Rocket*, the template for future locomotives, wins the Rainhill Trials.

▷ **1830** *Tom Thumb* and *The Best Friend of Charleston* herald the start of locomotive building in the US.

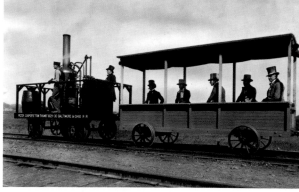

△ **Tom Thumb**
On the trial run of Peter Cooper's *Tom Thumb* in 1830 (re-enacted here), the locomotive hauled a car containing 18 directors of the B&O Railroad.

▷ **1834** The first railway in Ireland is opened between Dublin and Kingstown.

▷ **1835** Germany's first railway opens between Nuremberg and Fürth.

▷ **1835** Britain's Great Western Railway is incorporated.

Richard Trevithick
1771–1833

Although Richard Trevithick built the world's first working steam locomotives, his name is less familiar than that of other pioneers in British railway engineering. Trevithick's misfortune was that he invented many of his machines 20 years before the world was ready to use them. As well as his locomotives, Trevithick pursued other engineering projects, which included a paddle-wheeled barge, a steam hammer, a steam rolling mill, and a tunnel under the River Thames. Trevithick also spent 10 years in South America, where he used his steam engine designs to help open up silver mines in Peru. He returned to England in 1827, but died penniless six years later.

EVOLVING DESIGNS

Early on in his career, Trevithick worked as an engineer in the local mines in Cornwall. His familiarity with stationary engines used for winding and pulling meant that he was well placed to experiment with high-pressure engines or "strong steam", which offered greater pulling power for locomotives. He built his first steam vehicle in 1801, known as the *Puffer*. This ran on roads, not rails, but met an unfortunate end when it crashed into a house at Camborne in Cornwall and caught fire.

In 1803 Trevithick built a high-pressure engine that could run on iron rails. The Pen-y-Darren engine hauled 10 tons (10 tonnes) of iron and 70 passengers along a 9½-mile (15.3-km) iron rail track, proving its usefulness. Although the engine's weight eventually fractured the track, it was a milestone in the development of the locomotives.

Trevithick developed his third and final locomotive in 1808 and named it *Catch-me-who-can*. The train pulled passengers around a circular cast-iron track he had built in London. Eventually the weight of the train fractured the track and derailed the engine, but by then Trevithick had proven to the world that a steam locomotive could be run on tracks.

The money train
Catch-me-who-can (1809) was the first passenger train to charge a fare. People had to pay one shilling to ride the train, which travelled at just over 12 mph (19 km/h) around a circular demonstration track in London.

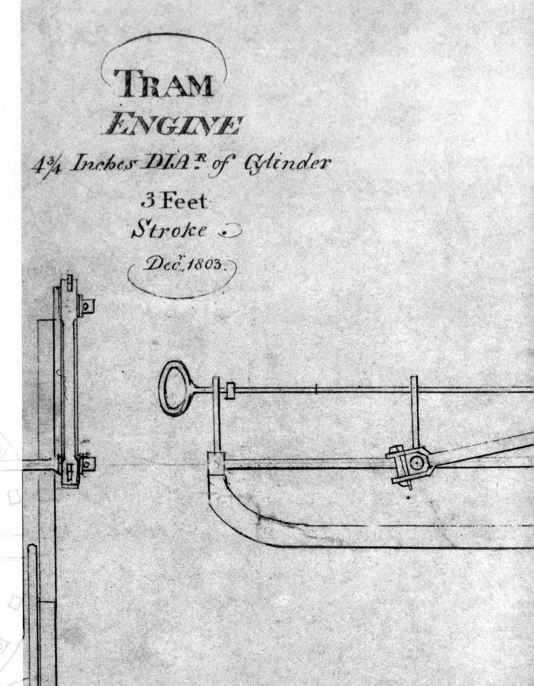

TRAM ENGINE
4¾ Inches DIA.ʳ of Cylinder
3 Feet Stroke
Dec.ʳ 1803.

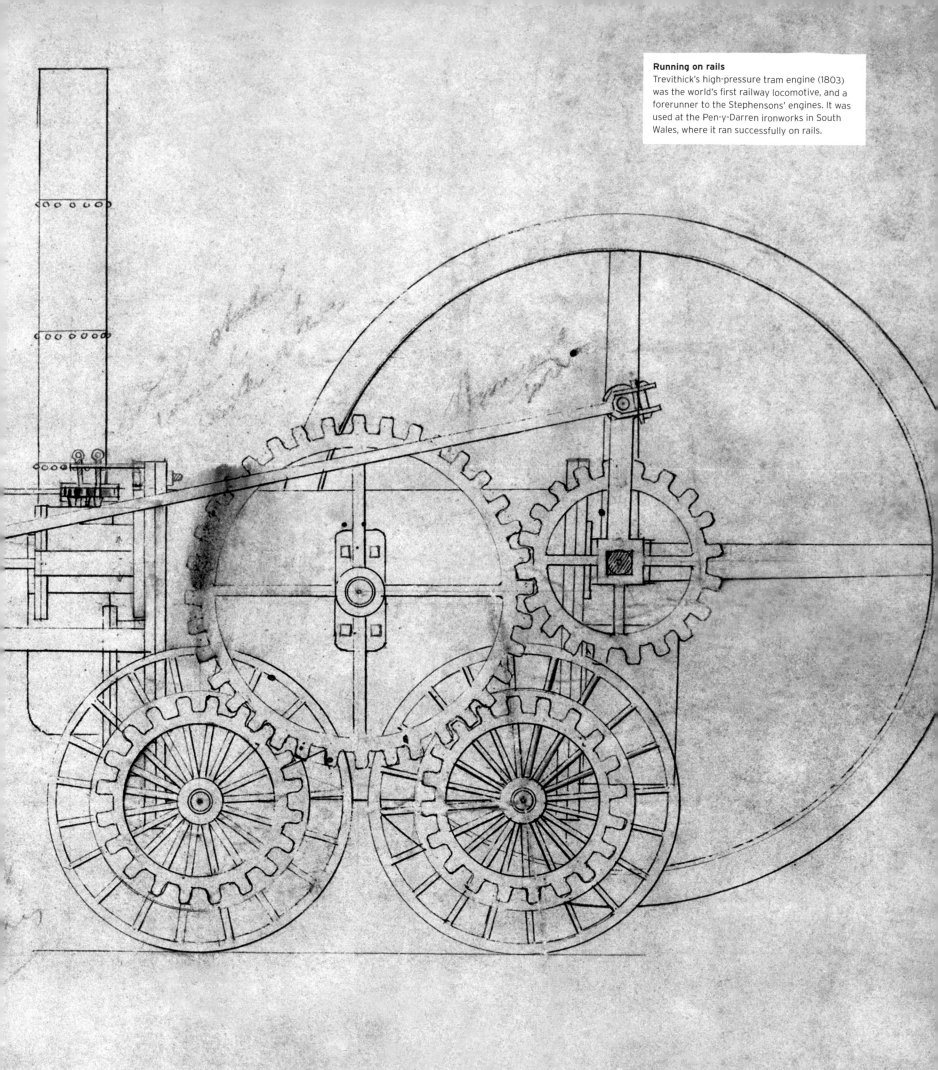

Running on rails
Trevithick's high-pressure tram engine (1803) was the world's first railway locomotive, and a forerunner to the Stephensons' engines. It was used at the Pen-y-Darren ironworks in South Wales, where it ran successfully on rails.

A British Invention

During the 18th century the British inventors Thomas Newcomen and James Watt led the way in the development of the low-pressure, stationary steam engines that played a vital role in the early years of the Industrial Revolution. A major breakthrough took place in the early 19th century when Cornish inventor Richard Trevithick successfully demonstrated the world's first working high-pressure, steam railway locomotive. From then on, British inventiveness, led by the "Father of Railways", George Stephenson, brought a rapid development, which culminated in 1830 with the opening of the world's first inter-city railway, between Liverpool and Manchester.

△ **Pen-y-darren locomotive, 1804**

Wheel arrangement	0-4-0
Cylinders	1
Boiler pressure	25 psi (1.75 kg/sq cm)
Driving wheel diameter	48 in (1,220 mm)
Top speed	approx. 5 mph (8 km/h)

Richard Trevithick's high-pressure steam locomotive hauled the world's first train on the Pen-y-darren Ironworks tramway in Merthyr Tydfil, South Wales on 13 February 1804. The train was carrying 10 ton (10.2 tonnes) of coal and 70 men.

◁ *Catch Me Who Can*, 1808

Wheel arrangement	2-2-0
Cylinders	1
Boiler pressure	25 psi (1.75 kg/sq cm)
Driving wheel diameter	48 in (1,220 mm)
Top speed	approx. 12 mph (19 km/h)

Richard Trevithick's *Catch Me Who Can* was demonstrated to the public on a circular track at a steam circus in Bloomsbury, London, in 1808. Unfortunately the train overturned when a rail broke, so the public was not convinced.

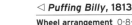

◁ *Puffing Billy*, 1813

Wheel arrangement	0-8-0 (final form 0-4-0)
Cylinders	2
Boiler pressure	40 psi (2.8 kg/sq cm)
Driving wheel diameter	48 in (1,220 mm)
Top speed	approx. 5 mph (8 km/h)

Weighing 7.25 tons (7.4 tonnes) and built by William Hedley for the Wylam Colliery in Northumberland, *Puffing Billy* was the world's first commercial adhesion steam engine. Now preserved at London's Science Museum, it is considered the oldest surviving locomotive.

▷ *Locomotion No. 1*, 1825

Wheel arrangement	0-4-0
Cylinders	2
Boiler pressure	50 psi (3.51 kg/sq cm)
Driving wheel diameter	48 in (1,220 mm)
Top speed	approx. 15 mph (24 km/h)

Built by George and Robert Stephenson, *Locomotion No.1* hauled the first train on the Stockton & Darlington Railway, the world's first public railway, in 1825. This locomotive has been preserved and can be seen at the Darlington Railway Museum, County Durham.

△ Rocket, 1829

Wheel arrangement 0-2-2

Cylinders 2

Boiler pressure 50 psi (3.51 kg/sq cm)

Driving wheel diameter 56³/₄ in
(1,435 mm)

Top speed approx. 30 mph (48 km/h)

Robert Stephenson & Co.'s advanced and innovative *Rocket* was the clear winner of the Rainhill Trials held on the Liverpool & Manchester Railway in 1829. The *Rocket* is shown pulling a first-class passenger carriage; luggage was carried on the roof.

▷ Agenoria, 1829

Wheel arrangement 0-4-0

Cylinders 2

Boiler pressure 40 psi (2.8 kg/sq cm)

Driving wheel diameter 48 in (1,220 mm)

Top speed approx. 8 mph (13 km/h)

One of only four steam locomotives built by Foster, Rastrick & Co. of Stourbridge, *Agenoria* worked on the Earl of Dudley's Shutt End Colliery Railway, Staffordshire, for 35 years. The same company built the *Stourbridge Lion*, the first locomotive to be exported to the US.

◁ Sans Pareil, 1829

Wheel arrangement 0-4-0

Cylinders 2

Boiler pressure 50 psi (3.51 kg/sq cm)

Driving wheel diameter 54 in (1,372 mm)

Top speed approx. 18 mph (29 km/h)

Built by Timothy Hackworth, *Sans Pareil* (meaning "without equal") performed well in the Rainhill Trials on the Liverpool & Manchester Railway in 1829 but exceeded the permitted weight, so was not considered for the prize.

▷ Novelty, 1829

Wheel arrangement 0-2-2WT

Cylinders 2

Boiler pressure 50 psi (3.51 kg/sq cm)

Driving wheel diameter 54 in (1,372 mm)

Top speed approx. 28 mph (45 km/h)

Although it was one of the fastest locomotives at the 1829 Rainhill Trials, John Ericsson and John Braithwaite's lightweight *Novelty* proved unreliable and was withdrawn. It was the first locomotive to have its cylinders within the frames.

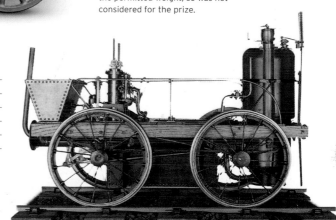

Rocket

The Rainhill Trials were staged in 1829 to decide which locomotives would run the world's first "inter-city" passenger trains on the Liverpool & Manchester Railway (L&MR) from 15 September 1830. Built by engineer Robert Stephenson, *Rocket* competed in the trials and hit a top speed of 28 mph (45 km/h). As the undisputed winner, *Rocket* clinched the prized contract, winning fame and universal acclaim for Stephenson.

ROCKET FEATURED A NUMBER of engineering innovations that ensured its success at the Rainhill Trials. It had inclined cylinders on either side of the firebox, which were connected to single driving wheels by short rods, giving it more thrust than could be achieved by the beam arrangement on earlier engines. It was the first engine to have a multitube boiler and chimney blastpipe, which greatly improved steam production. The basic design principles embodied in *Rocket* carried

through to the last steam locomotives. The original 1829 *Rocket* can be seen in London's Science Museum, but was extensively modified. The replica shown here is a more accurate representation of the original. A working replica built in 1979 for the 150th anniversary of the L&MR resides at the National Railway Museum in York. It incorporates the trailing and tender wheelsets and iron frame from a replica built at Crewe Works in 1880 to mark the centenary of George Stephenson's birth.

SPECIFICATIONS FOR ORIGINAL ROCKET	
Class	Rocket
Wheel arrangement	0-2-2
Origin	UK
Designer/builder	Robert Stephenson & Co.
Number produced	5
In-service period	1830–40
Cylinders	2, inclined at 37 degrees
Boiler pressure	50psi (3.51 kg/sq cm)
Driving wheel diameter	56³/₄ in (1,435 mm)
Top speed	approx. 30 mph (48 km/h)

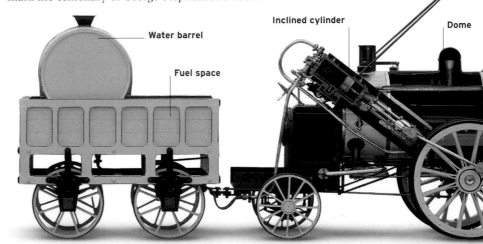

Chimney coronet

16 ft (4.9 m) chimney

Chimney stays

Inclined cylinder

Dome

Water barrel

Fuel space

CROSS SECTION OF *ROCKET* WITH TENDER (ABOVE)   FIRST CLASS CARRIAGES (BELOW)

Roof luggage rack

Six-seat compartment

Hand-painted company name

Sprung buffers with safety chain link

Carriage name

Oak-framed windows

Guard's seat

Revolutionary engine
This *Rocket* replica includes many of the features that facilitated the speeds the original achieved at the trials. *Rocket* was the first locomotive to have a fully functional blastpipe, which forced exhaust steam up the chimney. The engine had no brakes. Stopping was achieved via a foot pedal that puts the engine into reverse gear.

ROCKET.

SIDE VIEW

FRONT VIEW

ROCKET.

EXTERIOR

The 1979 working replica *Rocket* could never be wholly faithful to the original because of the need to conform to modern health and safety requirements and operating conditions. The replica was initially fitted with wooden driving wheels, which had to be replaced with steel ones when the originals buckled as the engine derailed on rough track at Bold Colliery in May 1980, forcing it to miss the opening of the *Rocket 150* celebrations. The replica's "water tank" was a 54-gallon (245-litre) "Hogshead" beer barrel originally from the Wadworth Brewery in Devizes, Wiltshire.

1. Boiler barrel brass nameplate   **2.** Chimney coronet   **3.** Pressure gauge shut-off valve
4. Boiler water-level test taps   **5.** Crosshead, showing connecting rod small end and piston rod
6. Timber driving wheel and connecting rod   **7.** Small end detail   **8.** Works plate   **9.** Left side
driving rod   **10.** Slip eccentric assembly   **11.** Water barrel   **12.** Tender wheel and safety chains

FOOTPLATE

Rocket has a small basic footplate that provides no weather protection for the crew. The "fallplate" (the metal plate that bridges the gap between engine and tender) slides and rocks about when in operation, so in wet and windy weather the driver can feel as if he is at sea. The propensity for dropped lumps of coke or coal to lodge beneath the floor-mounted, valve-gear treadle and firebox-damper handle could make driving conditions particularly difficult.

13. Firebox with copper, main steam-feed pipes above **14.** Valve gear operating treadle **15.** Regulator valve **16.** Firebox door **17.** Right-side, valve-gear control levers

CARRIAGES

The carriages seen with the *Rocket* replica are both reproductions of original 1834 L&MR first-class coaches, built in 1930 for the railway's centenary. The carriages each have three six-seat compartments, and are named *Traveller* and *Huskisson* – the latter after Liverpool MP William Huskisson, who was struck and killed by the locomotive at the L&MR's opening ceremony in 1830.

18. Carriage buffer **19.** Tender buffer spring **20.** Brass hand grip **21.** Carriage name in gold leaf **22.** Carriage steps for passengers **23.** Carriage wheel, axle box, and leaf spring **24.** Guard's seat **25.** Carriage window strap **26.** First-class "button back" upholstered seats

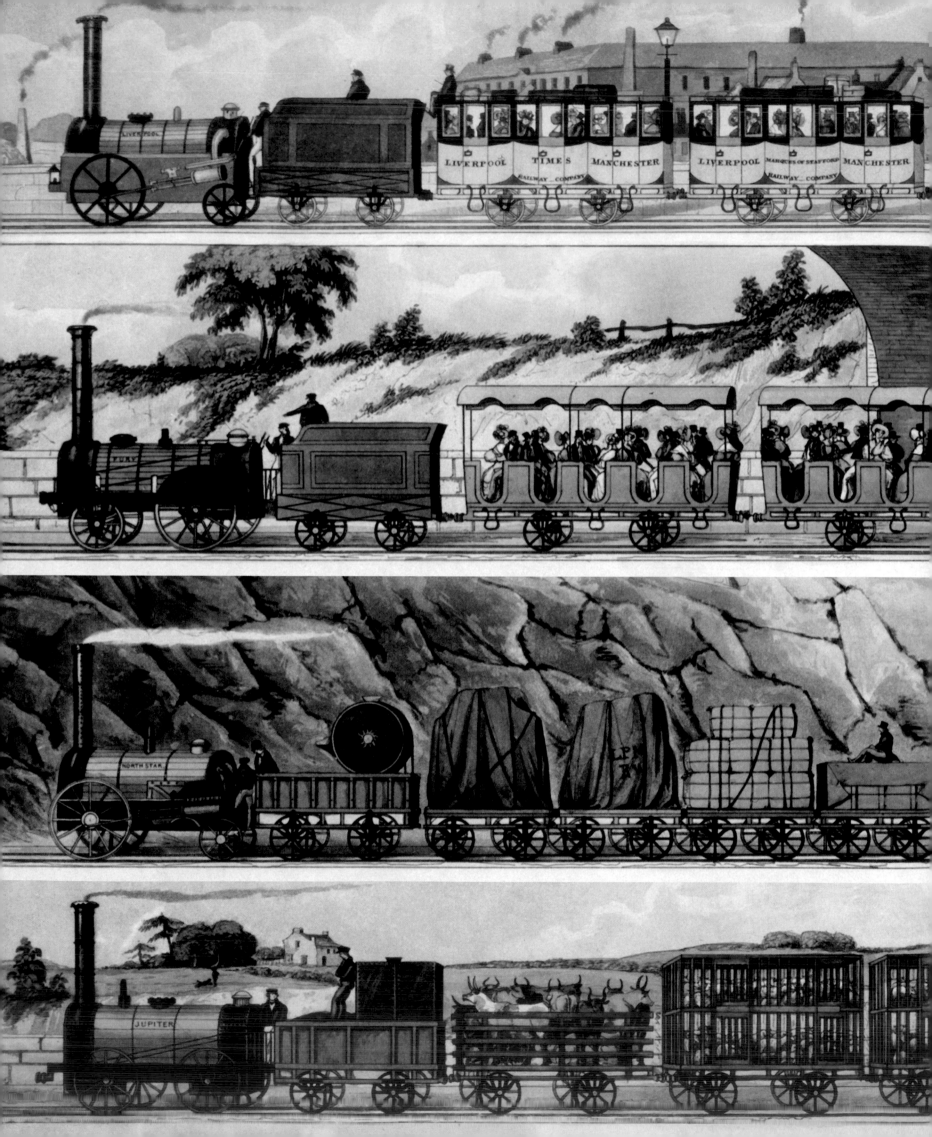

The Liverpool & Manchester Railway

The Liverpool & Manchester Railway (L&MR) opened in 1830 and was the world's first railway to carry both fare-paying passengers and freight. It established a cheaper and more efficient transport link between the factories of Manchester and Lancashire and the port of Liverpool. This delighted factory owners, who sought faster and cheaper routes than those provided by boats on the Bridgewater Canal.

ENGINEERING CHALLENGE

George Stephenson was appointed engineer of the twin-track, 32-mile (51.5-km) line. His expertise was put to the test at Chat Moss near Manchester, where the track crossed an unstable 4-mile (6.5-km) stretch of peat bog. The terrain almost brought the project to a halt, but Stephenson overcame the challenge by having the line built on a floating foundation of wood and stone.

Within three years, 64 bridges and viaducts had been constructed along the line, including the nine-arch Sankey Viaduct, the Wapping Tunnel in Liverpool, and a 2-mile (3.2-km) cutting through Olive Mount. A passenger terminus was also built at each end of the line in Manchester and Liverpool. In the first six months of 1831, the L&MR carried 188,726 passengers and almost 36,000 tons (36,578 tonnes) of freight.

From top to bottom: *Liverpool* with first-class carriages and a mail coach; *Fury* with second-class carriages; *North Star* pulling goods wagons; and *Jupiter* transporting livestock.

Steam for Home and Export

The success of Stephenson's *Rocket* and the opening of the world's first public railway in 1825 and the inter-city route in 1830 led to demand for British-built steam railway locomotives at home and abroad. The most successful of the early builders was Robert Stephenson & Company of Newcastle-upon-Tyne, founded by George and his son Robert in 1823. Its early locomotives were built for the Stockton & Darlington Railway but it also supplied locomotives for the first railways in Egypt and Germany as well as the US.

◁ *Invicta*, 1829-30

Wheel arrangement	0-4-0
Cylinders	2
Boiler pressure	40 psi (2.81 kg/sq cm)
Driving wheel diameter	48 in (1,220 mm)
Top speed	approx. 20 mph (32 km/h)

Robert Stephenson & Co. built *Invicta* in Newcastle, then shipped it to Kent (UK) by sea. *Invicta* hauled the first train on the Canterbury & Whitstable Railway in 1830. The locomotive was named after the motto "invicta" (undefeated) on the flag of Kent. It is on display at Kent's Canterbury Museum.

▽ *John Bull*, 1831

Wheel arrangement	0-4-0 (as built) 2-4-0 (as modified)
Cylinders	2 (inside)
Boiler pressure	45 psi (3.16 kg/sq cm)
Driving wheel diameter	66 in (1,676 mm)
Top speed	approx. 30 mph (48 km/h)

Built by Robert Stephenson & Co., *John Bull* was exported to the US, where it worked on the Camden & Amboy Railroad from 1831 to 1866. US engineer Isaac Dripps added his two-wheel bogie, to which he attached the first cowcatcher, as well as a headlight, spark-arresting chimney, and covered tender and cab.

▷ *Planet*, 1830

Wheel arrangement	2-2-0
Cylinders	2 (inside)
Boiler pressure	45 psi (3.16 kg/sq cm)
Driving wheel diameter	66 in (1,676 mm)
Top speed	approx. 35 mph (56 km/h)

Planet was the first type to have inside cylinders and the ninth locomotive built for the Liverpool & Manchester Railway. Designed by Robert Stephenson & Co., *Planet* was the first engine type to be built in large numbers.

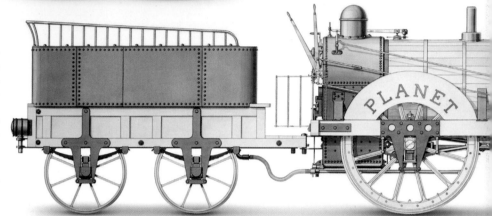

JOHN BULL **AS FIRST CONSTRUCTED, 1831**

▽ *Adler*, 1835

Wheel arrangement	2-2-2
Cylinders	2 (inside)
Boiler pressure	48 psi (3.37 kg/sq cm)
Driving wheel diameter	54 in (1,372 mm)
Top speed	approx. 17 mph (27 km/h)

The *Adler* (meaning "eagle") was the first successful steam railway locomotive to operate in Germany. It was built for the Bavarian Ludwig Railway by Robert Stephenson & Co. *Adler* remained in service until 1857. In 1935 a replica was built to mark the centenary of the German railways.

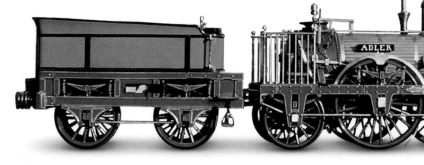

▽ **Bury**, 1831

Wheel arrangement	0-4-0
Cylinders	2 (inside)
Boiler pressure	50 psi (3.52 kg/sq cm)
Driving wheel diameter	66 in (1,676 mm)
Top speed	approx. 40 mph (64 km/h)

These locomotives were built with bar frames to reduce weight and were noted for their round-topped fireboxes. Designed by Edward Bury & Co., the Bury was popular in the US where light track was laid quickly to cover vast distances.

◁ *North Star*, 1838

Wheel arrangement	2-2-2
Cylinders	2 (inside)
Boiler pressure	50 psi (3.52 kg/sq cm)
Driving wheel diameter	84 in (2,134 mm)
Top speed	approx. 40 mph (64 km/h)

Robert Stephenson & Co.'s *North Star* hauled the inaugural director's train on the broad-gauge Great Western Railway in 1838. The locomotive was rebuilt in 1854 and withdrawn from service in 1871.

△ Hawthorn *Sunbeam*, 1837

Wheel arrangement	2-2-0
Cylinders	2 (inside)
Boiler pressure	50 psi (3.52 kg/sq cm)
Driving wheel diameter	60 in (1,524 mm)
Top speed	approx. 40 mph (64 km/h)

Sunbeam was built by R. & W. Hawthorn & Co. of Newcastle for the Stockton & Darlington Railway. Hawthorn built marine and stationary steam engines as well as locomotives for the broad-gauge Great Western Railway.

▷ *Lion*, 1838

Wheel arrangement	0-4-2
Cylinders	2 (inside)
Boiler pressure	50 psi (3.52 kg/sq cm)
Driving wheel diameter	60 in (1,524 mm)
Top speed	approx. 35 mph (56 km/h)

Lion was one of the first two locomotives built by Todd, Kitson & Laird. The other one was called *Tiger*. *Lion* worked on the Liverpool & Manchester Railway until 1859 before it was retired to Liverpool Docks as a stationary pumping engine.

The Stephensons
1781-1848/1803-59

GEORGE STEPHENSON
1781-1848

ROBERT STEPHENSON
1803-59

In 1830 the world's first passenger railway opened, the Liverpool & Manchester, heralding the dawn of mechanized transportation. The man responsible was George Stephenson, a self-taught colliery engineer, who is known as the "Father of the Railways" for his pioneering achievements in civil and mechanical engineering. Working with his engineer son Robert, Stephenson created a series of steam locomotives. The pair also collaborated on building the Stockton & Darlington Railway (1825), where George introduced his standard 4 ft 8½ in (1.44 m) rail gauge, which is still in use worldwide today.

A GROWING REPUTATION

George Stephenson was an innovator from the start. In 1814 he built his inaugural locomotive, *Blücher*, which was the first engine to use flanged wheels running on rails. In 1823 he set up a locomotive works in Newcastle with Robert that built the first steam engines to run on commercial railway lines. The company's first engine was named *Locomotion No. 1*, but perhaps the best known was *Rocket*, which serviced the Liverpool & Manchester Railway after winning a competition in 1829.

The Stephensons's growing reputation meant that they were much in demand as chief engineers to Britain's burgeoning rail network, following the Liverpool & Manchester with the London & Birmingham railway in 1833. They were even consulted on railway schemes overseas, in Egypt, Italy, and Norway. Robert's expertise also extended to railway bridges; he engineered the High Level Bridge in Newcastle (1849) and the Royal Border Bridge in Northumberland (1850), among others.

Digging deep
Approximately 480,000 cubic yards (367,000 cubic metres) of rock were excavated for the 2-mile- (3.2-km-) long Olive Mount Cutting on the Liverpool & Manchester Railway. The cutting is almost 70 ft (21m) deep in places.

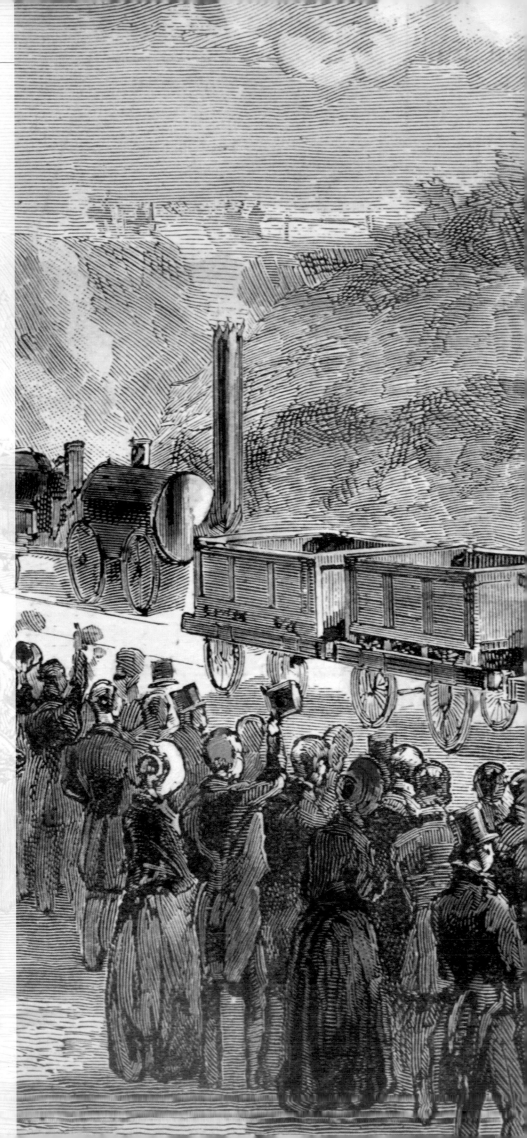

Winner takes all
Steam locomotive trials were run in October 1829 at Rainhill, near Liverpool, to decide which engine would be used on the Liverpool & Manchester Railway. Stephenson's *Rocket* triumphed, beating four other competitors.

World Pioneers

By the mid 1820s, pioneering inventors and engineers in continental Europe and the US were experimenting with their own designs. Some of these developments, such as US civil engineer John B. Jervis's swivelling leading bogie or Frenchman Marc Séguin's multitube boiler, would soon be incorporated into locomotives around the world. By the late 1830s rapid technological advances in steam locomotive design led to a massive expansion of railway building. In the US, the Baltimore & Ohio Railroad was the first to operate scheduled freight and passenger services. By 1837 the service had extended from Baltimore over the iconic Thomas Viaduct to Washington DC and across the Potomac River to Harper's Ferry.

▽ John Stevens's *Steam Waggon*, 1825

Wheel arrangement	early rack-and-pinion
Cylinders	1
Boiler pressure	approx. 100+ psi (7.03 kg/sq cm)
Driving wheel diameter	57 in (1,450 mm)
Top speed	approx. 12 mph (19 km/h)

Colonel John Stevens's *Steam Waggon* demonstrated the practicability of very high-pressure steam railway locomotives. This was the first engine to run on rails in the US. Stevens ran it on a circular track on his estate in Hoboken, New Jersey.

△ Marc Séguin's locomotive, 1829

Wheel arrangement	0-4-0
Cylinders	2
Boiler pressure	approx. 35 psi (2.46 kg/sq cm)
Driving wheel diameter	approx. 54 in (1,372 mm)
Top speed	approx. 15 mph (24 km/h)

Fitted with a multitube boiler, enormous rotary blowers, and a large firebox, Marc Séguin's innovative steam locomotive was the first to be built in France. It was tested on the Saint-Étienne & Lyon Railway in November 1829 and entered regular service in 1830.

259   The *Best Friend*, from William H. Brown, *The History of the First Locomotives in America*, New York, 1874

THE CHARLESTON & HAMBURG RAILROAD

△ *Best Friend of Charleston*, 1830

Wheel arrangement	0-4-0
Cylinders	2
Boiler pressure	approx. 35 psi (2.46 kg/sq cm)
Driving wheel diameter	approx. 57 in (1,450 mm)
Top speed	approx. 25 mph (40 km/h)

The first steam locomotive to be constructed entirely in the US, *Best Friend of Charleston* was built by the West Point Foundry in New York. It operated a passenger service on the South Carolina Railroad until it was destroyed by a boiler explosion.

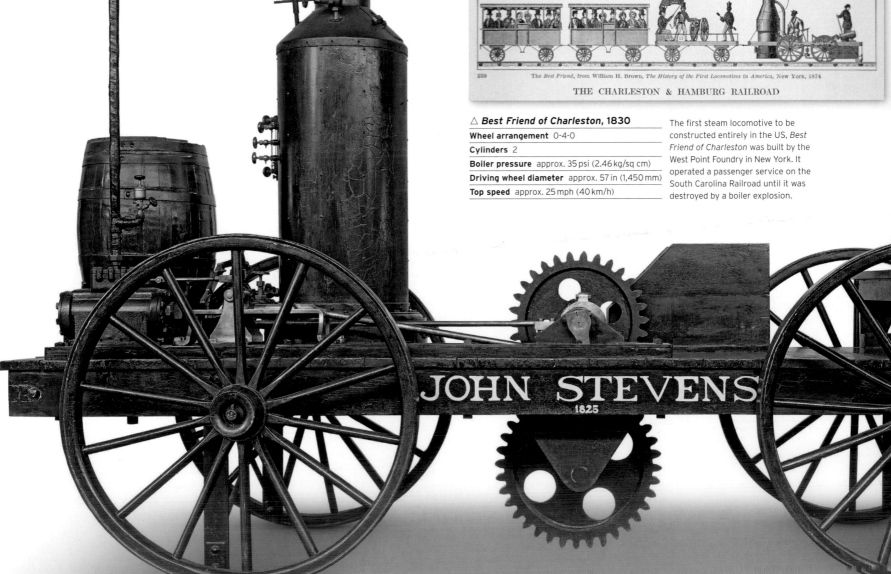

JOHN STEVENS
1825

PIONEERS

Marc Séguin, 1786–1875

Born in the Ardèche region of France, engineer, inventor, and entrepreneur Marc Séguin built innovative steam locomotives for the Saint-Étienne & Lyon Railway. His engines were fitted with an ingenious multi-tube boiler, which he patented in 1827, as well as mechanically driven fans to improve draughting for the fire and a firebox enclosed by a water jacket for greater heating capacity. Séguin developed the first suspension bridge in continental Europe and went on to build 186 bridges in France.

Engineering Innovation Marc Séguin was inspired by George Stephenson's *Locomotion No. 1*, which he saw in action on the Stockton & Darlington Railway in 1825.

◁ *Tom Thumb*, 1830

Wheel arrangement	2-2-0
Cylinders	1
Boiler pressure	approx. 35 psi (2.46 kg/sq cm)
Driving wheel diameter	approx. 33 in (840 mm)
Top speed	14 mph (23 km/h)

This locomotive was built by US inventor and, later, presidential candidate Peter Cooper. The Baltimore & Ohio Railroad raced *Tom Thumb* against a horse to decide whether they should adopt steam power or horse traction; the train lost, but the railroad saw its potential. Weighing only 1.1 ton (1 tonne), *Tom Thumb* had a vertical boiler with inner tubes fashioned from gun barrels.

▷ *DeWitt Clinton*, 1831

Wheel arrangement	0-4-0
Cylinders	2
Boiler pressure	approx. 35 psi (2.46 kg/sq cm)
Driving wheel diameter	approx. 58¾ in (1,520 mm)
Top speed	approx. 20 mph (32 km/h)

The first steam locomotive to operate in New York State, the *DeWitt Clinton* was built for the Mohawk & Hudson Railroad. Passengers travelled in converted stage coaches. It was named after a governor of New York State who was responsible for the construction of the Erie Canal.

◁ *Experiment*, 1832

Wheel arrangement	4-2-0
Cylinders	2
Boiler pressure	approx. 50 psi (3.51 kg/sq cm)
Driving wheel diameter	approx. 72 in (1,830 mm)
Top speed	approx. 60 mph (96 km/h)

This engine was designed by John B. Jervis, chief engineer for the Delaware & Hudson Canal & Railroad. *Experiment*, later named *Brother Jonathan*, was built by the West Point Foundry, New York, for use on the Mohawk & Hudson Railroad. It was the first locomotive with a leading bogie that became the 4-2-0 type.

Railroad Expansion

The earliest US railroads were operated using horse power. In 1830 the Baltimore & Ohio Railroad (B&O) was one of the first to introduce steam. While some railroads bought designs from fledging manufacturers such as Baldwin, the B&O started constructing their own, including the long-lived "Grasshoppers". In 1836 William Norris introduced the four-wheel leading bogie, which became common worldwide until the end of steam in the 20th century. Two years later Johann Schubert's *Saxonia* became the first successful steam engine to be built and operated in Germany.

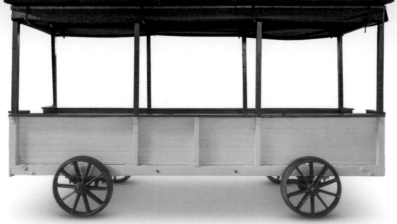

△ **Baldwin *Old Ironsides*, 1832**

Wheel arrangement	2-2-0
Cylinders	2
Boiler pressure	50 psi (3.51 kg/sq cm)
Driving wheel diameter	54 in (1,372 mm)
Top speed	approx. 28 mph (45 km/h)

Designed by US inventor Matthias Baldwin, *Old Ironsides* was the first commissioned steam locomotive built at the Baldwin Locomotive Works, for the Philadelphia, Germantown & Norristown Railroad.

▷ **B&O *Atlantic*, 1832**

Wheel arrangement	0-4-0
Cylinders	2
Boiler pressure	50 psi (3.52 kg/sq cm)
Driving wheel diameter	35 in (890 mm)
Top speed	approx. 20 mph (32 km/h)

Built by US inventor and foundry owner Phineas Davis for the Baltimore & Ohio Railroad, *Atlantic* was the prototype for 20 more similar locomotives nicknamed "Grasshoppers".

1832   BALTIMORE & OHIO R. R. ATLANTIC   1832

Early Coaches

The first railway passenger coaches in the US were primitive affairs, often based on existing designs for turnpike stagecoaches and originally intended for low-speed, horse-operated railroads. The rail companies soon learnt that they were impractical: seats were uncomfortable, passengers in open-air carriages not only had to brave the elements but also the smoke, hot ash, and cinders blown out by the equally primitive steam locomotives that hauled the coaches.

◁ **Director's Car, 1828**

Type	4-wheel
Capacity	12 passengers
Construction	iron and wood
Railway	Baltimore & Ohio Railroad

Originally horsedrawn, in August 1830 the Baltimore & Ohio Director's Car carried the railroad's directors in the first steam-hauled train along the railway to Ellicott's Mills behind *Tom Thumb*. This is a replica built in 1926 for the Fair of the Iron Horse and can be seen at the B&O Railroad Museum, Baltimore.

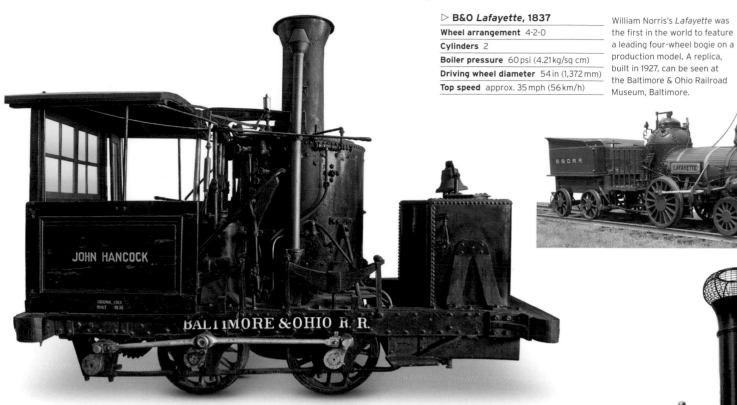

▷ B&O *Lafayette*, 1837

Wheel arrangement	4-2-0
Cylinders	2
Boiler pressure	60 psi (4.21 kg/sq cm)
Driving wheel diameter	54 in (1,372 mm)
Top speed	approx. 35 mph (56 km/h)

William Norris's *Lafayette* was the first in the world to feature a leading four-wheel bogie on a production model. A replica, built in 1927, can be seen at the Baltimore & Ohio Railroad Museum, Baltimore.

△ B&O "Grasshopper" *John Hancock*, 1836

Wheel arrangement	0-4-0
Cylinders	2
Boiler pressure	50 psi (3.51 kg/sq cm)
Driving wheel diameter	35 in (889 mm)
Top speed	approx. 20 mph (32 km/h)

Fitted with a driver's cab, *John Hancock* was one of 20 "Grasshopper" locomotives built by the Baltimore & Ohio Railroad. It remained in service as a switcher until 1892.

▷ *Saxonia*, 1838

Wheel arrangement	0-4-2
Cylinders	2
Boiler pressure	60 psi (4.21 kg/sq cm)
Driving wheel diameter	59 in (1,500 mm)
Top speed	approx. 37 mph (60 km/h)

Designed by Johann Schubert, *Saxonia* was the first practical working steam locomotive built entirely in Germany. It was used on the Leipzig to Dresden Railway –Germany's first long-distance line. By 1843 *Saxonia* had clocked up more than 5,300 miles (8,500 km).

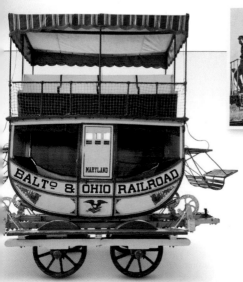

◁ Maryland Coach, 1830

Type	4-wheel
Capacity	14 passengers
Construction	iron and wood
Railway	Baltimore & Ohio Railroad

Based on a stagecoach, Richard Imlay's double-deck coach was one of six built for the inaugural steam train on the Baltimore & Ohio Railroad. The carriage body was perched on unsprung wheels cradled on leather straps. It was unstable and offered little protection for top deck passengers.

▷ Nova Scotia Coach, 1838

Type	4-wheel
Capacity	6 passengers
Construction	iron and wood
Railway	General Mining Association of Nova Scotia

Built by Timothy Hackworth of London (UK), the Nova Scotia Coach carried the Director of Nova Scotia's General Mining Association on the colliery railway on Cape Breton Island in Canada. Also known as the bride's car, it was said to have originally carried the director's new bride to their home after their marriage ceremony.

1839-1869
BUILDING
NATIONS

BUILDING NATIONS

On 10 May 1869 a "golden spike" was hammered into a sleeper in a dusty part of Utah at Promontory in the US – and two locomotives eased gently towards each other. The simple ceremony marked the completion of the first transcontinental railroad, and was a key moment in the development of the US. In the mid-19th century railways were to become the driving force of progress not just in the US, but throughout the world. Tracks spread across Europe and into ever more inaccessible places. In India, a country that would become one of the greatest railway nations, the first passenger train left Bombay in April 1853.

Yet this was still a time of experiments. Engineering genius Isambard Kingdom Brunel built Britain's Great Western Railway (GWR) not to the normal 4 ft 8½ in (1.435 m) track gauge – but to his own much wider 7 ft ¼ in (2.14 m). The bigger gauge allowed for high speeds and more spacious trains, but too much track had already been laid to the narrower width favoured by Stephenson. Mediating in the "gauge war", the UK Parliament decided against Brunel's idea.

Other inventions had long-lasting effects: the telegraph and "mechanical interlocking" that connected signals and points were developed through the 1850s, and in 1869 George Westinghouse introduced air brakes – now standard around the world. As railways grew, so did their reach through society; in 1842 Britain's Queen Victoria took her first train journey, travelling via the GWR on her way to Windsor. A fundamental change came with the birth of mass city transit when London's first underground line opened in 1863. As the network developed, labourers could travel cheaply to work from the city's outskirts, fuelling the creation of the world's first metropolis.

△ **India's first passenger train**
On 16 April 1853, GIP No.1 carried passengers from Bombay (now Mumbai) to Tanneh (now Thane) on the Great Indian Peninsular Railway.

Key Events

▷ **1839** Germany's first long-distance line opens, linking the cities of Leipzig and Dresden.

▷ **1839** In the Netherlands, Amsterdam and Haarlem are connected by the country's first railway.

▷ **1841** Thomas Cook invents the "charter" train, for a group of temperance campaigners travelling from Leicester to Loughborough (UK).

▷ **1842** Britain's Queen Victoria gives royal approval by travelling on the Great Western Railway.

▷ **1844** The railway reaches Basel in Switzerland, via France; Switzerland's first domestic line opens in 1847.

▷ **1850s** Safety is improved with mechanical interlocking that connects points and signals together.

▷ **1853** India's debut passenger train runs from Bombay to Thane.

▷ **1863** The world's first true underground railway – London's Metropolitan Railway – is opened.

△ **The Metropolitan Line**
Steam locomotives in tunnels meant that passengers had to contend with smoky stations and carriages that were lit by gas lamps.

▷ **1869** North America's first transcontinental railroad is completed.

▷ **1869** George Westinghouse of the US invents the air brake.

▷ **1869** George Mortimer Pullman launches the ultimate in luxury travel – the Pullman Car.

"Let the country **but make the railroads, and the railroads will make the country**"
EDWARD PEASE, BRITISH RAILWAY PROMOTER

◁ **The "golden spike" at Promontory, Utah, in 1869** linked the Union Pacific and Central Pacific railroads in the US

The US Forges Ahead

The British locomotives imported into the US were often too heavy for the lighter, quickly laid rail tracks, and not powerful enough to cope with the steeper gradients. So US engineers developed designs tailored to their railways' needs. A leading truck, first with two wheels then four, was fitted to guide the engines through the many sharp curves. Improved traction led to the 4-4-0 becoming the standard type, soon followed by the more powerful 4-6-0. Cowcatchers, headlights, and warning bells were fitted to cope with the unfenced tracks. American designers built locomotives capable of hauling heavy loads over a railroad system that by 1871 linked two oceans.

▷ B&O L Class No. 57 *Memnon*, 1848

Wheel arrangement	0-8-0
Cylinders	2
Boiler pressure	75 psi (5 kg/sq cm)
Driving wheel diameter	44 in (1,118 mm)
Top speed	approx. 30 mph (48 km/h)

Bought by the Baltimore & Ohio Railroad in 1848 for freight working, *Memnon* was later used in the Civil War, for hauling troops and supplies. Eight driving wheels gave this locomotive its extra power and traction.

▽ CVR No. 13 *Pioneer*, 1851

Wheel arrangement	2-2-2
Cylinders	2
Boiler pressure	100 psi (7 kg/sq cm)
Driving wheel diameter	54 in (1,372 mm)
Top speed	approx. 40 mph (64 km/h)

Pioneer hauled the short passenger trains of the Cumberland Valley Road of Pennsylvania and western Maryland until 1890. It survived the destruction of the railway's workshops by the Confederate troops in 1862.

TALKING POINT

Financing the Railroads

Railroad promoters looked to the commercial centres of Philadelphia, Boston, and New York, as well as European money markets to raise capital to develop the railways. Investors preferred bonds to stocks since these offered a guaranteed income. At the same time, the US government offered federal land grants to the rail companies, who then sold the land they did not need to raise more funds.

B&O stocks The value of shares in the US's new Baltimore & Ohio Railroad exceeded $3 million in 1839 when this $100 certificate was issued.

▷ W&A No. 39 *The General*, 1855

Wheel arrangement	4-4-0
Cylinders	2
Boiler pressure	140 psi (10 kg/sq cm)
Driving wheel diameter	60 in (1,524 mm)
Top speed	approx. 45 mph (72 km/h)

Built by the Western & Atlantic Railroad, *The General* pulled passenger and freight trains between Atlanta, Georgia, and Chattanooga, Tennessee, from 1856 until 1891.

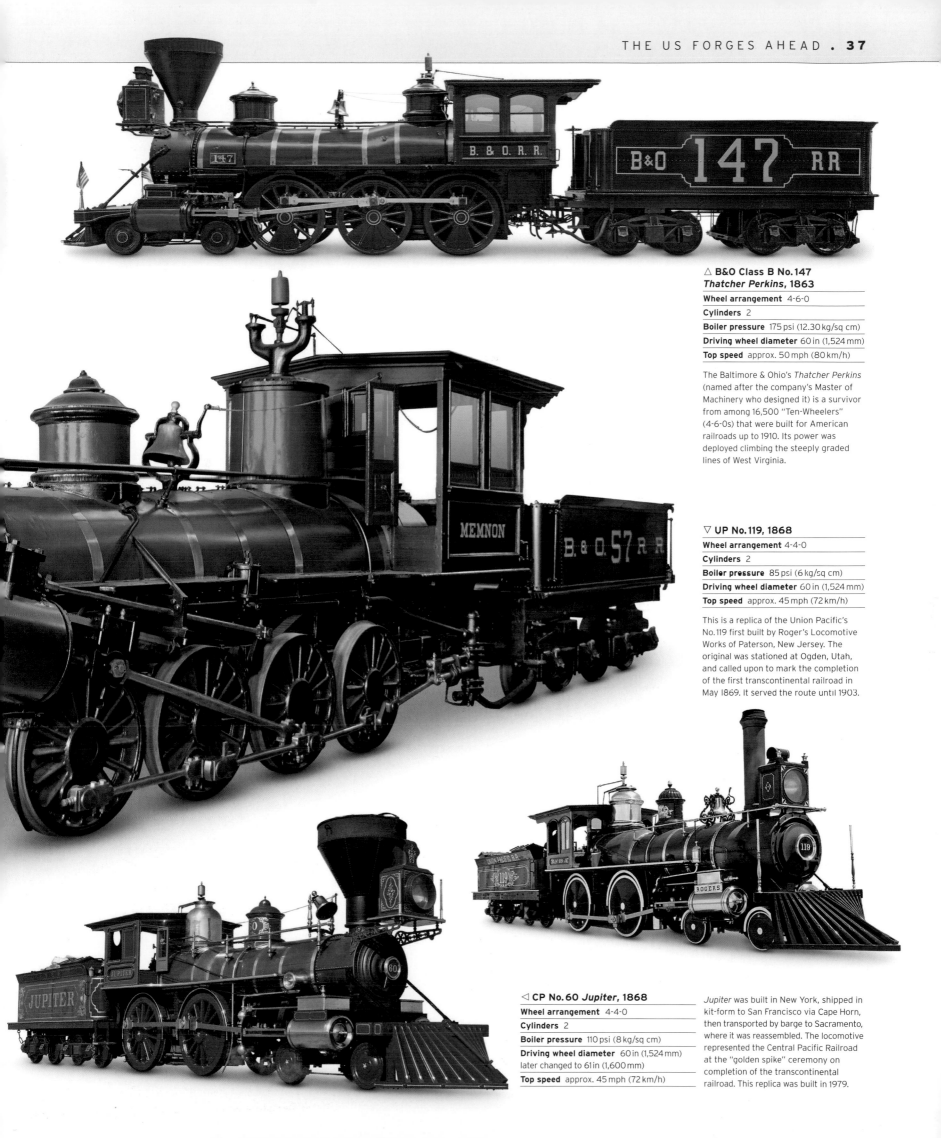

△ **B&O Class B No.147**
***Thatcher Perkins*, 1863**

Wheel arrangement	4-6-0
Cylinders	2
Boiler pressure	175 psi (12.30 kg/sq cm)
Driving wheel diameter	60 in (1,524 mm)
Top speed	approx. 50 mph (80 km/h)

The Baltimore & Ohio's *Thatcher Perkins* (named after the company's Master of Machinery who designed it) is a survivor from among 16,500 "Ten-Wheelers" (4-6-0s) that were built for American railroads up to 1910. Its power was deployed climbing the steeply graded lines of West Virginia.

▽ **UP No.119, 1868**

Wheel arrangement	4-4-0
Cylinders	2
Boiler pressure	85 psi (6 kg/sq cm)
Driving wheel diameter	60 in (1,524 mm)
Top speed	approx. 45 mph (72 km/h)

This is a replica of the Union Pacific's No.119 first built by Roger's Locomotive Works of Paterson, New Jersey. The original was stationed at Ogden, Utah, and called upon to mark the completion of the first transcontinental railroad in May 1869. It served the route until 1903.

◁ **CP No.60 *Jupiter*, 1868**

Wheel arrangement	4-4-0
Cylinders	2
Boiler pressure	110 psi (8 kg/sq cm)
Driving wheel diameter	60 in (1,524 mm) later changed to 61 in (1,600 mm)
Top speed	approx. 45 mph (72 km/h)

Jupiter was built in New York, shipped in kit-form to San Francisco via Cape Horn, then transported by barge to Sacramento, where it was reassembled. The locomotive represented the Central Pacific Railroad at the "golden spike" ceremony on completion of the transcontinental railroad. This replica was built in 1979.

Thatcher Perkins

Designed by the Master of Machinery at the Baltimore & Ohio (B&O) Railroad, Thatcher Perkins, the Class B No. 147 was built in 1863. It entered service the same year, and was used to transport Union troops during the American Civil War. Subsequently, No. 147 hauled freight and passenger trains in West Virginia until its retirement in 1892. It was given the name *Thatcher Perkins* for the B&O's centennial celebrations in 1927.

WITH EXTRA GRIP from its 4-6-0 wheel arrangement, No. 147 was a natural progression from the 4-4-0 locomotive workhorses first used by US railways. Fitted with Stephenson link-motion valve gear and a large, spark-arresting smokestack and oil lamp, this 40-ton (41-tonne) locomotive was designed to pull first-class passenger trains on the company's steeply graded line from Cumberland to Grafton in what is now West Virginia. It replaced a similar locomotive destroyed in the American Civil War in 1861, and began service hauling Union troops and munitions across the Allegheny Mountains during the war.

The locomotive's heavy build kept it in service for 29 years, after which it was retired and preserved by the B&O for exhibitions and other public-relations purposes. Since 1953 *Thatcher Perkins* has been on display in the Mount Clare Roundhouse at the B&O Railroad Museum in Baltimore. However, in 2003 the building's roof collapsed during a blizzard and the locomotive was seriously damaged. It has since been restored, and is now back on display in the museum.

FRONT VIEW

REAR VIEW

Leading the way
Opening in 1827, the Baltimore & Ohio Railroad was the first railway in the US to operate scheduled freight and passenger services for the public.

SPECIFICATIONS			
Class	B	**In-service period**	1863–92 (Thatcher Perkins)
Wheel arrangement	4-6-0	**Cylinders**	2
Origin	USA	**Boiler pressure**	175 psi (12.3 kg/sq cm)
Designer/builder	Thatcher Perkins/B&O	**Driving wheel diameter**	60 in (1,524 mm) as built
Number produced	11 Class B	**Top speed**	approx. 50 mph (80 km/h)

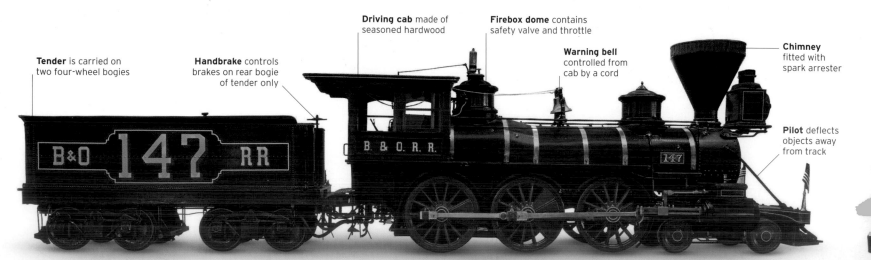

Driving cab made of seasoned hardwood

Firebox dome contains safety valve and throttle

Warning bell controlled from cab by a cord

Chimney fitted with spark arrester

Tender is carried on two four-wheel bogies

Handbrake controls brakes on rear bogie of tender only

Pilot deflects objects away from track

Safety first
When first introduced No.147 burned enormous quantities of wood, which it carried in the eight-wheel tender. The locomotive's spark-arresting chimney was fitted with a double layer of mesh that stopped wood embers floating away and setting fire to railside buildings and vegetation.

EXTERIOR

The B&O painted No. 147 in a bold colour scheme. The large headlamp, pilot (cowcatcher), sand dome, driving cab, and tender body were finished in Indian red with gold lettering and lining. The locomotive and tender wheels as well as the top of the smokestack were painted in vermilion, while the cylinders, smoke box, chimney, wheel splashers, under parts, and boiler casing were black. Finally, the boiler bands, flag holders, bell, and oil cups were made of brass. Unlike in Europe, it was common practice in North America to fit pilots, or cowcatchers, to the front of steam locomotives to deflect obstacles from the track.

1. Engine number plate **2.** Pilot (cowcatcher) **3.** Headlight **4.** Oil cup for lubricating steam chest **5.** Cylinder housing piston **6.** Linkage for valve gear **7.** Brass bell with decorative mounting yoke **8.** Sand dome **9.** Brass whistle **10.** Oil cups for lubricating side rods **11.** Driving wheels and side rods **12.** Tender bogie (truck) **13.** Crosshead **14.** Cab windows **15.** Steps up to tender **16.** Link-and-pin coupler at rear of tender

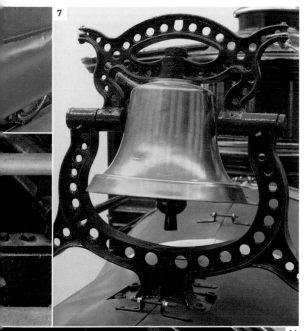

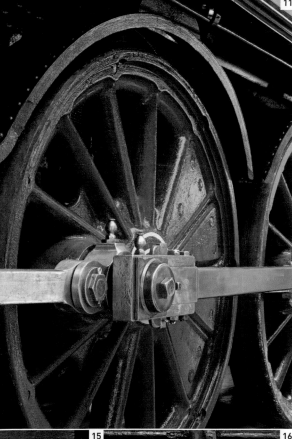

CAB AND TENDER

The spacious driving cab was built of wood and protected the driver and fireman from the elements; the cab was also fitted with arched windows, allowing a good view of the track ahead. Cords to operate the whistle and bell hung from the roof, while seats were arranged at each side of the firebox door and offered the crew a touch of comfort.

Early American steam locomotives used vast quantities of wood carried in a large tender at the rear. No. 147's tender, which also contained a water tank taking up the two sides and rear, was carried on two four-wheel bogies.

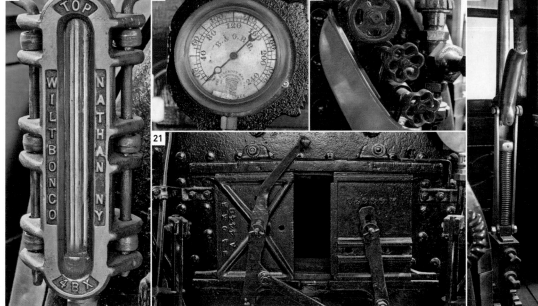

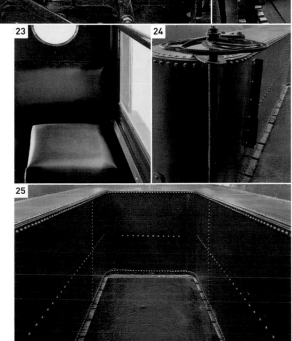

17. Locomotive cab (rear view) 18. Water-level gauge (sight glass) 19. Boiler pressure gauge 20. Water tri-cocks 21. Firebox doors 22. Reverser bar (Johnson bar) 23. Fireman's seat 24. Handbrake wheel 25. Tender coal bunker

Building Great Railways
Union Pacific

Completed in 1872, the first transcontinental railway across North America linked Chicago with California. Today the route is owned by Union Pacific, North America's largest Class 1 freight railway.

Union Pacific freight train
The Union Pacific owns nearly 95,000 freight cars and operates double-stack intermodal freight over its 31,800 miles (51,177 km) of track between the West and East coasts.

THE UNION PACIFIC RAILROAD (UP) started life in 1862 when President Lincoln signed the Pacific Railroad Act authorizing the building of the first transcontinental railway across North America. Following the wagon-train trails made by pioneer emigrants heading west, the UP was to build westwards from Omaha on the west bank of the Missouri River, Nebraska. The Central Pacific Railroad (CP) was to build eastwards from Sacramento, California.

The CP began laying track from Sacramento in 1863. All of the railway equipment for this section had first to be brought on a long and often dangerous voyage around Cape Horn from the East Coast, a journey that could sometimes take several months.

Travel poster
A woman overlooks a lush Californian valley in this promotional poster for Union Pacific from around 1915.

Led by the railway's first General Manager, Thomas Durant, and with a workforce of Irish navvies, the UP commenced building westwards along the Platte River Valley from Omaha in 1865. Railway equipment was first delivered for the UP by riverboats. However, the opening of the Chicago, Iowa & Nebraska Railroad (later, the Chicago & North Western Railroad) linking Chicago to Council Bluffs on the

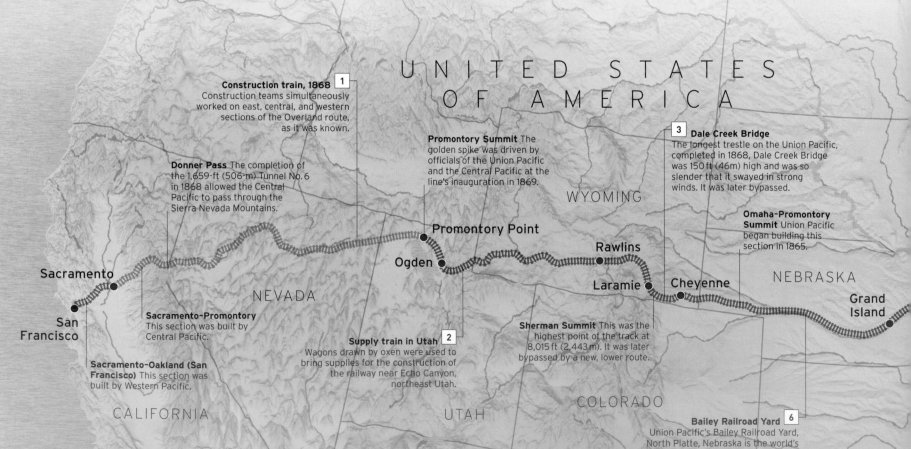

UNITED STATES OF AMERICA

Construction train, 1868 [1]
Construction teams simultaneously worked on east, central, and western sections of the Overland route, as it was known.

Donner Pass The completion of the 1,659-ft (506-m) Tunnel No. 6 in 1868 allowed the Central Pacific to pass through the Sierra Nevada Mountains.

Promontory Summit The golden spike was driven by officials of the Union Pacific and the Central Pacific at the line's inauguration in 1869.

Dale Creek Bridge [3]
The longest trestle on the Union Pacific, completed in 1868, Dale Creek Bridge was 150 ft (46 m) high and was so slender that it swayed in strong winds. It was later bypassed.

Omaha–Promontory Summit Union Pacific began building this section in 1865.

WYOMING

Promontory Point

Ogden

Rawlins

NEVADA

Laramie

Cheyenne

NEBRASKA

Sacramento

Grand Island

San Francisco

Sacramento–Promontory This section was built by Central Pacific.

Supply train in Utah [2]
Wagons drawn by oxen were used to bring supplies for the construction of the railway near Echo Canyon, northeast Utah.

Sherman Summit This was the highest point of the track at 8,015 ft (2,443 m). It was later bypassed by a new, lower route.

Sacramento–Oakland (San Francisco) This section was built by Western Pacific.

COLORADO

CALIFORNIA

UTAH

Bailey Railroad Yard [6]
Union Pacific's Bailey Railroad Yard, North Platte, Nebraska is the world's largest marshalling yard

Union Pacific diesels
The Union Pacific owns just over 8,000 diesel-electric locomotives, one of which is seen here at the head of a freight train on the Overland route through the California desert.

east bank of the Missouri River opposite Omaha, allowed materials to be delivered by train, and by 1868 this section had reached Sherman Summit.

Meanwhile, in the west, the CP employed 12,000 Chinese labourers to construct 15 tunnels to reach Donner Pass by 1868.

The two railways met at Promontory on 9 May 1869, where ceremonial golden spikes were driven into the final wooden sleeper or "tie". However, the transcontinental railway was only finally completed in 1872 when the Union Pacific opened up its bridge across the Missouri River, linking Omaha and Council Bluffs.

Passenger trains were discontinued in 1971 when the newly formed Amtrak took over responsibility for these services. Today, apart from a daily passenger service aboard the luxurious *California Zephyr*, the route carries only freight.

KEY FACTS

DATES
1863 First Central Pacific rails laid at Sacramento
1865 First Union Pacific rails laid in Omaha
1869 Golden spike ceremony at Promontory
1872 Missouri River Bridge completes the line
1883 First passenger service on *Overland Flyer*

TRAINS
First steam locomotive 4-4-0 *Major General Sherman* built in 1864, first saw service in 1865
Largest steam locomotive 4000-Class 4-8-8-4 articulated locomotives or "Big Boys", 1941–44
Diesel-electric locomotives Union Pacific currently operates 8,000, including General Electric 4,400 hp CC44AC/CTE; EMD 4,000 hp SD70M

JOURNEY
Chicago to San Francisco 2,300 miles (3,700 km)
1893 *Overland Flyer* takes 86 hours 30 minutes including a ferry from Oakland to San Francisco
1906 The *Overland Limited* takes 56 hours
Current journey 51 hours

RAILWAY
Gauge 4 ft 8 ½ in (1,435 mm)
Tunnels Union Pacific: 4; Central Pacific: 15; longest is Summit Tunnel 1,750 ft (533 m)
Longest bridge Dale Creek Bridge 600 ft (183 m)
Highest point Sherman Summit 8,015 ft (2,443 m). Now bypassed

KEY
● Start/Finish
● Main stations
‖‖‖‖ Union Pacific
‖‖‖‖ Central Pacific
‖‖‖‖ Chicago, Iowa & Nebraska
‖‖‖‖ Western Pacific

4 The Overland train
The *Overland Flyer*, later the *Overland Limited*, ran part of the Union Pacific route from 1887. Passengers could enjoy the scenery from the observation car at the train's rear.

Chicago-Council Bluffs (Omaha)
This section of the line, built by Chicago, Iowa & Nebraska Railroad, predated the other sections of the Union Pacific Railroad.

IOWA

Chicago

5 *Forty-Niner* to San Francisco
The *Forty-Niner*, named for the miners of the California Gold Rush in 1849, was a heavyweight steam streamliner which departed five times a month from Chicago in the 1940s.

Omaha

ILLINOIS

Council Bluffs A ferry transferred passengers across the Missouri River to Omaha before the bridge was built.

N

0		150		300 miles
0	150	200	450 km	

A JOINT EFFORT

Although named the Union Pacific Railroad, the transatlantic route was originally built by four companies: the Chicago, Iowa & Nebraska Railroad; the Union Pacific; the Central Pacific; and the Western Pacific.

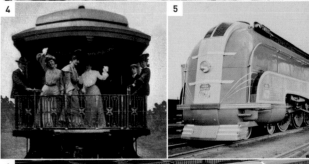

Britain Advances

This period of British railway history features both successes and failures. The Grand Junction Railway's famous Crewe Works opened on a green-field site in 1840 and was soon turning out graceful, single-wheeler express locomotives. While in Liverpool, Edward Bury pursued his bar-frame design, which became popular in North America. On the downside, Brunel's atmospheric railway in Devon was an unmitigated disaster, and the failure of John Fowler's underground steam locomotive caused the designer much embarrassment.

△ **FR No. 3 "Old Coppernob", 1846**

Wheel arrangement	0-4-0
Cylinders	2 (inside)
Boiler pressure	100 psi (7 kg/sq cm)
Driving wheel diameter	57 in (1,448 mm)
Top speed	approx. 30 mph (48 km/h)

Nicknamed "Old Coppernob" because of the copper cladding around its firebox, this locomotive was designed by Edward Bury, and built at Bury, Curtis & Kennedy of Liverpool for the Furness Railway in northwest England. It is normally at the National Railway Museum, York, and is the only survivor of the bar-frame design in the UK.

△ **Fireless locomotive "Fowler's Ghost", 1861**

Wheel arrangement	2-4-0
Cylinders	2 (inside)
Boiler pressure	160 psi (11.25 kg/sq cm)
Driving wheel diameter	72 in (1,830 mm)
Top speed	approx. 20 mph (32 km/h)

This experimental locomotive, designed by John Fowler and built by Robert Stephenson & Co., was intended for use on London's broad-gauge Metropolitan underground railway. The engine was fitted with condensing apparatus to prevent steam and smoke emissions; it was a complete failure.

▷ **GJR Columbine, 1845**

Wheel arrangement	2-2-2
Cylinders	2
Boiler pressure	120 psi (8.43 kg/sq cm)
Driving wheel diameter	72 in (1,830 mm)
Top speed	approx. 40 mph (64 km/h)

The locomotive Columbine, designed by Alexander Allen, was the first to be built at the Grand Junction Railway's Crewe Works. It was subsequently used to haul the London & North Western Railway's Engineering Department Inspection Saloon. It hauled passenger trains until 1877 and was withdrawn in 1902. It is now a static exhibit at London's Science Museum.

◁ **FR *Prince*, 1863**

Wheel arrangement	0-4-0ST
Cylinders	2
Boiler pressure	160 psi (11.25 kg/sq cm)
Driving wheel diameter	24 in (610 mm)
Top speed	approx. 20 mph (32 km/h)

Businessman and engineer George England designed and built *Prince*. It was one of the first three steam locomotives delivered to the slate-carrying 1-ft 11¹/₂-in- (0.60-m-) gauge Ffestiniog Railway in North Wales in 1863. It was returned to service in 2013 for the 150th anniversary of steam on the railway, and is the line's oldest working engine.

▷ **LSWR Class 0298, 1863**

Wheel arrangement	2-4-0WT
Cylinders	2
Boiler pressure	160 psi (11.25 kg/sq cm)
Driving wheel diameter	67 in (1,702 mm)
Top speed	approx. 40 mph (64 km/h)

The Class 0298 was designed by Joseph Beattie for the London & South Western Railway to provide suburban passenger services in southwest London. A total of 85 of these well-tank locomotives were built, the majority by Beyer Peacock & Co.

Brunel's Atmospheric Railway

British engineer Isambard Kingdom Brunel built the broad-gauge South Devon Railway between Exeter and Totnes as an "atmospheric" railway. Dispensing with locomotives, trains were pushed along by a long piston enclosed in a cast-iron tube in the middle of the track. The vacuum to move the piston was created at stationary pumping houses (such as the one above). The railway opened in 1847, but failed within a year. In 1848 it was converted to operate with conventional haulage, because the grease that was applied to the leather flap that sealed the pipe melted during hot weather, or was eaten by rats.

Atmospheric railway track This section of Brunel's broad-gauge track with its cast-iron vacuum pipe is on display at Didcot Railway Centre.

◁ **LNWR *Pet*, 1865**

Wheel arrangement	0-4-0ST
Cylinders	2 (inside)
Boiler pressure	120 psi (8.43 kg/sq cm)
Driving wheel diameter	15 in (380 mm)
Top speed	approx. 5 mph (8 km/h)

John Ramsbottom, the locomotive superintendent of the London & North Western Railway, designed this engine. *Pet* is a small cabless steam locomotive that worked on the 1-ft 6-in- (0.45-m-) narrow-gauge Crewe Works Railway until 1929. It is now a static exhibit at the National Railway Museum, York.

Euro Progress

The 1840s saw rapid railway building across Europe, with many locomotive designs still heavily influenced by British engineering expertise; many had set up workshops in France and Austria. By the 1850s Thomas Crampton's unusual long-boilered, "single-wheeler" engines were hauling trains between Paris and Strasbourg at speeds exceeding 70 mph (113 km/h). The design and craftsmanship of locomotives built by fledgling European builders such as Strauss of Munich stood the test of time with many remaining in service well into the 20th century.

▷ **Oldenburgische Class G1 No.1** *Landwührden*, **1867**

Wheel arrangement	0-4-0
Cylinders	2
Boiler pressure	142 psi (9.98 kg/sq cm)
Driving wheel diameter	59 in (1,500 mm)
Top speed	37 mph (60 km/h)

The first locomotive to be built by Georg Krauss of Munich, No.1 *Landwührden* won a gold medal for excellence and design of workmanship at the World Exhibition in Paris in 1867. After first working on the Grand Duchy of Oldenburg State Railways' branch lines this lightweight engine was retired in 1900 and is now on display at the Deutsches Museum in Munich.

▽ **SNB** *Limatt*, **1847**

Wheel arrangement	4-2-0
Cylinders	2
Boiler pressure	85 psi (6 kg/sq cm)
Driving wheel diameter	59 in (1,500 mm)
Top speed	approx. 35 mph (56 km/h)

Built by Emil Kessler of Karlsruhe, Germany, *Limatt* was the first steam locomotive on the Swiss Northern Railway (Schweizerische Nordbahn, or SNB), Switzerland's first railway. The engine is named after the River Limmat, which the railway followed for much of its route. It is on display at the Swiss Museum of Transport in Luzern.

▷ **CF de l'Est Crampton, 1852**

Wheel arrangement	4-2-0
Cylinders	2
Boiler pressure	120 psi (0.43 kg/sq cm)
Driving wheel diameter	84 in (2,134 mm)
Top speed	79 mph (127 km/h)

These fast locomotives, designed by British engineer Thomas Crampton, featured a large driving wheel at the rear and a low mounted boiler. Built by Jean-Francois Cail, No.80 *Le Continent* hauled express trains between Paris and Strasbourg, retiring only in 1914 after covering 1.5 million miles (2.4 million km).

◁ **Südbahn Class 23 GKB 671, 1860**

Wheel arrangement	0-6-0
Cylinders	2
Boiler pressure	98 psi (6.89 kg/sq cm)
Driving wheel diameter	49 in (1,245 mm)
Top speed	28 mph (45 km/h)

This engine was built by the Lokomotivfabrik der StEG of Vienna to haul freight trains on the Graz Köflacher Railway in southern Austria. Still used to haul excursion trains, GKB 671 is the oldest steam locomotive in continuous use in the world.

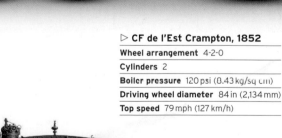

△ CF de l'Ouest Buddicom Type 111 No.33 Saint-Pierre, 1844

Wheel arrangement	2-2-2
Cylinders	2
Boiler pressure	80 psi (5.62 kg/sq cm)
Driving wheel diameter	75 in (1,905 mm)
Top speed	37 mph (60 km/h)

Built in Rouen, France, by British engineer William Buddicom for the new Paris to Rouen railway, No.33 Saint-Pierre had a long and successful career, retiring only in 1912. It is the oldest original steam locomotive still preserved on the European mainland, and is on display at the Cité du Train Museum in Mulhouse.

◁ BG Type 1B N2T Muldenthal, 1861

Wheel arrangement	2-4-0
Cylinders	2
Boiler pressure	110 psi (8 kg/sq cm)
Driving wheel diameter	48 in (1,220 mm)
Top speed	30 mph (48 km/h)

Sächsische Maschhinenfabrik of Chemnitz built the Type 1B N2T Muldenthal to haul coal trains on the newly opened Bockwaer Railway in Saxony. When retired in 1952 it was the oldest operational locomotive in Germany. It is now on display at the Dresden Transport Museum.

TALKING POINT

Class Travel

From the very early days, rail passengers were sorted according to their ability to pay and their position in society. While first-class passengers got sumptuous seating and plenty of space, second class was often very overcrowded and the seats were generally wooden. Those in third class travelled in uncovered wagons open to the elements, and to the smoke, cinders, and ash from the steam engine at the front.

FIRST CLASS

SECOND CLASS

THIRD CLASS

A Day at the Races, 1846 This cartoon from the London Illustrated News shows the social distinctions of class travel on the railways in Britain.

Isambard Kingdom Brunel 1806-59

Audacious and controversial, Isambard Kingdom Brunel became Britain's most innovative and successful engineer of the Victorian era. As a young man he worked on designs for bridges and commercial docks with his father Marc, an emigré French inventor and engineer. Brunel's career took off in 1826 when he was appointed resident engineer for the Thames Tunnel scheme between Wapping and Rotherhithe in London. Besides designing several of Britain's most famous railways, bridges, viaducts, and tunnels, Brunel was also involved in several dock schemes and three designs for transatlantic ships, which together transformed the face of Victorian England.

GREAT WESTERN RAILWAY

Brunel's most enduring contribution to railway development in Britain was the Great Western Railway (GWR) linking Bristol with London. Despite having no previous experience of railway engineering, he was selected for what was the most technically challenging civil engineering project of its time.

Building began simultaneously from both the London and Bristol termini in 1835, and the line opened in 1841. The 118-mile (190-km) route became famous for its smooth ride, and earned the GWR the nickname "Brunel's billiard table". Brunel's desire to establish the GWR as the fastest and most comfortable line saw him adopt a broad rail gauge of 7 ft ¼ in (2.14 m) instead of George Stephenson's standard gauge of 4 ft 8½ in (1.44 m), which had been used on railway lines in the Midlands and the North. It led to the "Battle of the Gauges", which lasted 50 years until the GWR finally embraced the standard gauge in 1892. The line remains a key route on Britain's rail network.

Brunel's final engineering project was the Royal Albert Bridge, famous for its gigantic tubular arches. By the time it was complete in 1859, Brunel was too ill to attend the opening, but managed to view his imposing masterpiece by lying on a platform truck that was hauled slowly across the bridge.

Box Tunnel
Built in 1841, Box Tunnel in Wiltshire linked the final section of the GWR between Chippenham and Bath. The construction of the 2-mile- (3.2-km-) long tunnel claimed the lives of more than 100 labourers.

Reaching Cornwall
Brunel built the Royal Albert Bridge (1859)
to carry the railway across the River Tamar into
Cornwall, extending the network westwards.
Here a prefabricated bridge span is being
prepared to be raised into position.

The GWR's Broad Gauge

While other British railways were being built to the standard gauge of 4 ft 8½ in (1.435 m), engineer Isambard Kingdom Brunel used the broad gauge of 7 ft ¼ in (2.14 m) when building the Great Western Railway, which opened from London Paddington to Bristol in 1841. Brunel had argued that his design offered higher speeds, smoother running, more stability, and increased comfort for passengers when compared to standard-gauge railways. In many ways he was right, but the spread of the standard gauge not only in Britain but also in many other parts of the world, including North America, led to Brunel's broad gauge becoming an anachronism. The GWR's last broad-gauge train ran on 21 May 1892.

◁ **GWR Firefly Class**
Fire Fly, 1840

Wheel arrangement	4-2-2
Cylinders	2 (inside)
Boiler pressure	100 psi (7 kg/sq cm)
Driving wheel diameter	84 in (2,134 mm)
Top speed	approx. 58 mph (93 km/h)

Designed by Daniel Gooch, Firefly was one of 61 express passenger locomotives built for the Great Western Railway by various builders between 1840 and 1842. The class was known for its speed with the original *Fire Fly* travelling from Twyford to Paddington in only 37 minutes. Built in 2005, this working replica is the 63rd *Fire Fly*. It operates at Didcot Railway Centre.

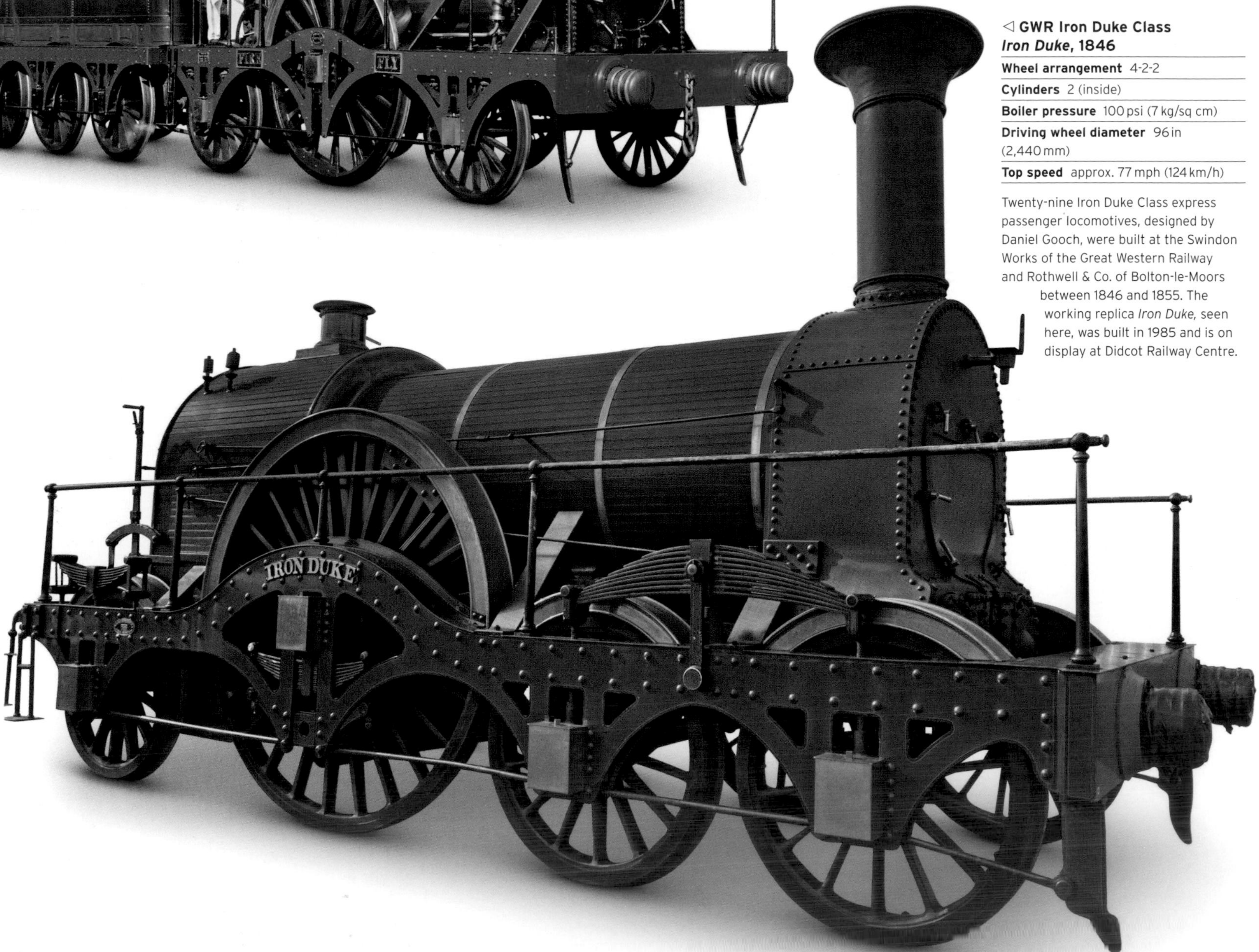

◁ **GWR Iron Duke Class**
Iron Duke, 1846

Wheel arrangement	4-2-2
Cylinders	2 (inside)
Boiler pressure	100 psi (7 kg/sq cm)
Driving wheel diameter	96 in (2,440 mm)
Top speed	approx. 77 mph (124 km/h)

Twenty-nine Iron Duke Class express passenger locomotives, designed by Daniel Gooch, were built at the Swindon Works of the Great Western Railway and Rothwell & Co. of Bolton-le-Moors between 1846 and 1855. The working replica *Iron Duke*, seen here, was built in 1985 and is on display at Didcot Railway Centre.

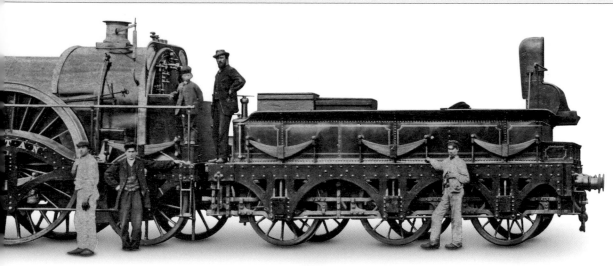

◁ **GWR Iron Duke Class *Sultan*, 1857**

Wheel arrangement	4-2-2
Cylinders	2 (inside)
Boiler pressure	100 psi (7 kg/sq cm)
Driving wheel diameter	96 in (2,440 mm)
Top speed	approx. 77 mph (124 km/h)

One of the Great Western Railway's Iron Duke Class express locomotives, *Sultan* was originally built in 1847, but was involved in an accident at Shrivenham a year later when it ran into a goods train. The prototype of this class, *Great Western*, was originally fitted with one pair of carrying wheels at the front as a 2-2-2. As with other members of the class, *Sultan*'s driving wheels had no flanges to allow movement on curves.

▷ **GWR Iron Duke Class *Lord of the Isles*, 1851**

Wheel arrangement	4-2-2
Cylinders	2 (inside)
Boiler pressure	140 psi (10 kg/sq cm)
Driving wheel diameter	96 in (2,440 mm)
Top speed	approx. 77 mph (124 km/h)

Another express passenger locomotive designed by Daniel Gooch for the Great Western Railway, *Lord of the Isles* was an improved version of the Iron Duke Class with higher boiler pressure, sanding gear, and a better driver's "cab". When new, it was exhibited at the Great Exhibition of 1851, and then in Chicago in 1893. It was withdrawn in 1884.

◁ **GWR Rover Class, 1870/1871**

Wheel arrangement	4-2-2
Cylinders	2 (inside)
Boiler pressure	145 psi (10.19 kg/sq cm)
Driving wheel diameter	96 in (2,440 mm)
Top speed	approx. 77 mph (124 km/h)

Built between 1871 and 1888, the Great Western Railway's Rover Class of express locomotives was similar to the Iron Duke Class, but with a small increase in boiler pressure and more protective driver's cabs. They used names previously carried by Iron Dukes and stayed in service until the end of the broad gauge in 1892.

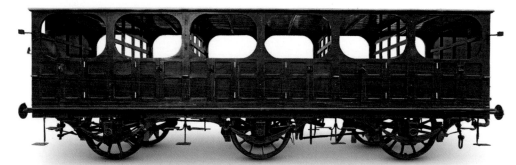

△ **GWR Broad Gauge Coach, 1840**

Type	6-wheel, Second Class
Capacity	48 passengers
Construction	iron chassis, wooden coach body
Railway	Great Western Railway

This replica of a Great Western Railway, broad-gauge, second-class carriage was built by London's Science Museum to run with their replica *Iron Duke* locomotive, to celebrate the anniversary of the railway in 1985. It now operates with *Fire Fly* at Didcot Railway Centre.

TECHNOLOGY

Battle of the Gauges

There were major problems for passengers who were forced to change trains at stations where the Great Western Railway's broad gauge met standard-gauge tracks. In 1846 the British Government passed the Railway Regulation (Gauge) Act, which mandated the 4-ft 8½-in (1.435-m) gauge for UK and 5 ft 3 in (1.6 m) for Ireland. Brunel was overruled, and by 1892 all the GWR's lines were converted to standard gauge.

***Break of Gauge at Gloucester*, 1846** This political cartoon depicts the confusion caused at Gloucester station where passengers with luggage had to change trains from the broad-gauge Great Western Railway to the standard-gauge Midland Railway and vice versa.

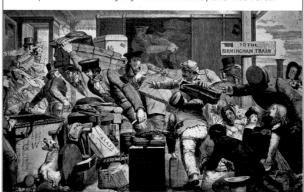

Mass Movers

As railways expanded, so did their roles and with that the need for engines designed for specific purposes. Express passenger engines had large driving wheels, which increased the distance travelled in each rotation. For goods trains, haulage power was transmitted through six, eight, or ten smaller wheels that provided the adhesion necessary for trains to move heavy loads. Suburban passenger services kept to timetables by using tank engines that could run equally well smokebox- or bunker-first. For branch line and shunting engines, size and weight were key factors, so the short wheelbase 0-4-0 and the 2-4-0 and 0-6-0 types were preferred.

◁ S&DR No. 25 *Derwent*, 1845

Wheel arrangement	0-6-0
Cylinders	2
Boiler pressure	75 psi (3.5 kg/sq cm)
Driving wheel diameter	48 in (1,220 mm)
Top speed	approx. 10-15 mph (16-24 km/h)

From the middle of the 19th century, the six-wheel goods engine became the principal British locomotive. One of the earliest, Timothy Hackworth's *Derwent* of 1845, served the Stockton & Darlington Railway, in northeast England, until 1869.

▷ Met Class A No. 23, 1864

Wheel arrangement	4-4-0T
Cylinders	2
Boiler pressure	120.13 psi (8.46 kg/sq cm); later 150 psi (10.53 kg/sq cm)
Driving wheel diameter	60½ in (1,537 mm)
Top speed	approx. 45 mph (72 km/h)

Tank locomotives built by Beyer Peacock of Manchester were the mainstay of London's Metropolitan Railway from the 1860s until the advent of electrification. To cut pollution, exhaust steam was returned to the water tanks where it was condensed for reuse.

Wagons and Carriages

Unsurprisingly, the designs of the earliest railway vehicles were based on proven ideas. Carriages adopted the design of the road coach; wagons were no more than enlarged versions of the iron and wooden, four-wheel tubs that had been used in mines for centuries. However, increasing loads – both passenger and goods – faster speeds, and the call for greater comfort and facilities brought about rapid advances.

△ SH Chaldron Wagon, 1845-55

Type	Bucket-type coal wagon
Weight	3⅓ tons (3.35 tonnes)
Construction	Iron platework and chassis
Railway	South Hetton Colliery

The design of the chaldron – a medieval measure used for weighing coal – was adopted for the earliest type of wagon. This one was used on George Stephenson's railway at the South Hetton Colliery, County Durham, which opened in 1822.

◁ LNWR "Large Bloomers", 1851

Wheel arrangement	2-2-2
Cylinders	2
Boiler pressure	100 psi (7 kg/sq cm); later 150 psi (10.53 kg/sq cm)
Driving wheel diameter	84 in (2,134 mm)
Top speed	approx. 50–60 mph (80–96 km/h)

Designed by James McConnell, 74 of these single-wheeler passenger engines were built for the London & North Western Railway up to 1862. They mainly worked between London and Birmingham. The nickname, "Large Bloomers", is attributed to American reformer Amelia Bloomer who scandalized Victorian society by wearing trousers.

▷ S&PR No. 5 *Shannon*, 1857

Wheel arrangement	0-4-0WT
Cylinders	2
Boiler pressure	120 psi (8.43 kg/sq cm)
Driving wheel diameter	35 in (889 mm)
Top speed	approx. 10–12 mph (16–19 km/h)

London's George England & Co. built this well tank for the Sandy & Potton Railway in Bedfordshire. In 1862 *Shannon* was sold to the London & North Western Railway, spending 16 years as a works shunter before ending its career on the Wantage Tramway in Oxfordshire.

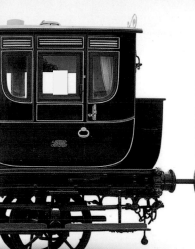

◁ L&BR Queen Adelaide's Saloon No. 2, 1842

Type	Passenger carriage with fold-down beds
Capacity	10 passengers
Construction	Wooden body, iron chassis
Railway	London & Birmingham Railway

This "stagecoach on wheels" transported Adelaide, Queen Consort to Britain's William IV. While the chassis was entrusted to the London & Birmingham Railway's Euston Works, the body was the work of a London coach builder. This is the oldest preserved carriage in Europe and is in the National Railway Museum, York.

△ NBR Dandy Car No. 1, 1863

Type	Horse-drawn rail car
Capacity	30 passengers (12 first and second class, 18 third class)
Construction	Wooden body and frame
Railway	North British Railway

Between 1863 and 1914 passengers on the Port Carlisle Railway in northwest England travelled in this horse-drawn Dandy Car, the horse trotting between the rails. First- and second-class passengers sat inside, while third class sat on benches at either end.

Building the Tube

Congestion on London's roads was a problem even during the mid-19th century. Charles Pearson, a city solicitor, decided to tackle the issue and was instrumental in raising the £1.3 million required to build the world's first underground railway line, the 3.75-mile- (6-km-) long Metropolitan. The line would link the City of London in the east and the Great Western Railway's terminus at Paddington to the west, with intermediate stations serving King's Cross and Euston.

Construction of the line alternated between open cuttings and tunnels, the latter mostly formed using a "cut-and-cover" method. This involved removing the street surface, cutting a trench, installing the retaining walls, track, and tunnel roof, and finally relaying the street surface.

INSTANT POPULARITY

Londoners immediately took to the underground line. On the opening day, 10 January 1863, 38,000 people rode in wooden-bodied, gas-lit carriages pulled by steam locomotives. Although their exhaust made conditions in stations unpleasant, it did not deter 9.5 million people from using the service in the first year.

The Metropolitan expanded 50 miles (80 km) to the north, but in London the future lay with deep-level lines, electric power, and narrower tunnels – what would become known as the Tube. The first deep-level Tube line opened with electric trains in 1890.

The tunnelling shield bore through soft, unstable soil such as clay during the excavation process. It acted as a barrier and support while spoil was removed.

Nations and Colonies

The success of the early British railways and steam engines attracted interest from across Europe and North America. As a result the newly industrialized countries such as US, France, and Germany began to lay the foundations for their own national systems, so became less and less dependent on British expertise.

However, Britain had a wider sphere of influence: its empire – the first railway outside Europe being built in the British colony of Jamaica. There were both economic and political reasons for the British to build railways in Australia, Canada, South Africa, and elsewhere. The vastness of India was controlled through its railway system, while the efficiency, and therefore profitability, of its mining, logging, and agriculture was completely transformed by the new transport.

DIE ERSTE BORSIG-LOKOMOTIVE
AUS DEM JAHRE 1841

△ Borsig No. 1, 1840

Wheel arrangement	4-2-2
Cylinders	2
Boiler pressure	80 psi (5.62 kg/sq cm)
Driving wheel diameter	54 in (1,372 mm)
Top speed	approx. 40 mph (64 km/h)

August Borsig opened a factory in Berlin in 1837 and three years later delivered his first locomotive to the Berlin-Potsdam Railway. In 1840, No.1 outpaced a British-built competitor, ending Germany's reliance on imports and helping make Borsig one of the world's leading engine builders.

△ I-class No. 1, 1855

Wheel arrangement	0-4-2
Cylinders	2
Boiler pressure	120 psi (8.43 kg/sq cm)
Driving wheel diameter	66 in (1,676 mm)
Top speed	approx. 20 mph (32 km/h)

One of four I-Class locomotives built by Robert Stephenson & Co of Newcastle-upon-Tyne, England, No.1 was delivered to the Sydney Railway Co. in January 1855. Train services were inaugurated in Australia that May. No.1 was retired in 1877 having run 156,542 miles (250,467 km).

▷ EIR No. 22 *Fairy Queen*, 1855

Wheel arrangement	2-2-2
Cylinders	2
Boiler pressure	80-100 psi (5.62-7 kg/sq cm)
Driving wheel diameter	72 in (1,830 mm)
Top speed	approx. 25 mph (40 km/h)

One of the first locomotives to haul passenger trains in India, *Fairy Queen* was built by Kitson, Hewitson & Thompson of Leeds, England, for the East Indian Railway. An outside-cylinder, 2-2-2 well tank, it is part of the historic locomotive collection in New Delhi and has a claim to be the world's oldest working engine.

△ *La Porteña*, 1857

Wheel arrangement	0-4-0ST
Cylinders	2
Boiler pressure	140-160 psi (9.84-11.25 kg/sq cm)
Driving wheel diameter	about 48 in (1,219 mm)
Top speed	approx 16 mph (26 km/h)

Arriving in Argentina from Britain on Christmas Day, 1856, the outside-cylindered, four-wheel saddletank *La Porteña* hauled the first train over the Buenos Aires Western Railway on 29 August 1857. Built by E.B. Wilson of Leeds, it remained in service until 1899 and is now exhibited at the museum in Luján.

◁ **Hawthorn No. 9 *Blackie*, 1859**

Wheel arrangement	0-4-2
Cylinders	2
Boiler pressure	130 psi (9.14 kg/sq cm)
Driving wheel diameter	54 in (1,372 mm)
Top speed	approx. 30 mph (48 km/h)

Hawthorn & Co. assembled this 0-4-0 at its works in Leith, Scotland, for contractor Edward Pickering, who used it in the construction of the 45-mile (72-km) Cape Town to Wellington Railway. South Africa's first locomotive, it was rebuilt as an 0-4-2 in 1873–74 and is now exhibited at Cape Town's main station.

△ **O&RR Class B No. 26, 1870**

Wheel arrangement	0-6-0
Cylinders	2
Boiler pressure	160–180 psi (11.25–12.65 kg/sq cm)
Driving wheel diameter	52 in (1,320 mm)
Top speed	approx. 40 mph (64 km/h)

This locomotive was built by Sharp, Stewart & Co. of Manchester, England, for the 5-ft 6-in- (1.67-m-) gauge Oudh & Rohilkhand Railway of northern India. No. 26 is typical of British engines exported at the time.

TECHNOLOGY

Challenging Railways

With mountain ranges, deserts, and jungles to be overcome, India posed a huge challenge to railway builders. Nevertheless, the first 25-mile (40-km) stretch between Bombay (now Mumbai) and Thane opened in November 1852, and by 1880 around 9,000 miles (14,484 km) of track had been laid. Twenty years on, the network had extended to 40,000 miles (64,374 km). A committee set up by the Governor General, Lord Dalhousie, led to the setting up of the Great Indian Peninsular Railway, the East India Railway, and the Darjeeling Himalayan Railway.

Construction site Workers photographed in 1856 on the wooden staging used in the building of the viaduct at the mouth of tunnel No. 8 (out of 28) on the Bhor Ghat Railway.

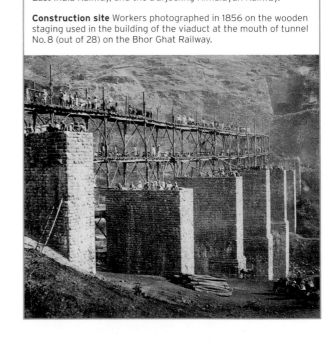

1870–1894

A WORLD
OF STEAM

A WORLD OF STEAM

When Bombay's Victoria Terminus opened in 1888, it was heralded as one of the world's grandest stations. Owing its styling to elements from both Indian and British history, it had taken 10 years to build. "VT" – now known as Mumbai's Chhatrapati Shivaji Terminus – became symbolic of an era in which nothing seemed beyond human endeavour and ingenuity.

At this time railways were spreading across the globe; they were climbing or boring through mountains and crossing mighty waterways via bridges, or being linked by steamships across vast seas and oceans. In 1881 the narrow-gauge Darjeeling Himalayan Railway opened, running from India's plains high into the foothills of the Himalayas. Meanwhile, the construction of Switzerland's Gotthard Tunnel had pushed a main line through 9 miles (15 km) of mountain rock. In 1885 the Canadian Pacific Railway was completed, creating a second route that spanned an entire continent. In the UK, the Forth Bridge opened in 1890 – crossing the Firth of Forth for more than 1½ miles (2.5 km). Then, in 1891, work started on a project that would dwarf almost everything else: Russia's Trans-Siberian Railway would join Moscow to Vladivostok on the country's far eastern coast.

There was an insatiable demand for more lines, higher speeds, more luxury, and greater magnificence. The railways' glamorous and luxurious side was epitomized by the development of the long-distance *Orient Express*, which by 1891 had connected Paris and Constantinople (Istanbul) via some of Europe's most important cities. Yet among all the expansions and improvements to steam travel, there were also early signs of a different future: in 1879 a new electric locomotive, which drew power from the track, was demonstrated in Berlin.

△ **Rush hour on the "El"**
In the late 19th century the Manhattan Railway Co. operated four elevated lines in New York City.

"**Lay down your rails**, ye Nations, near and far; Yoke your full trains to **Steam's triumphal car**"
CHARLES MACKAY, SCOTTISH POET

◁ **The US navy demonstrates steam** to Japanese onlookers in Yokohama in the late 19th century

Key Events

▷ **1870s** The electric "track circuit" is developed, which automatically shows signallers the location of trains.

▷ **1871** New York's Grand Central Station opens – it is later rebuilt as Grand Central Terminal.

▷ **1872** Japan's first railway opens between Tokyo and Yokohama.

▷ **1879** Werner von Siemens demonstrates an electric locomotive in Berlin; the following year an electric tramway is trialled in St Petersburg.

▷ **1881** The narrow-gauge Darjeeling Himalayan Railway is completed, connecting the Darjeeling hill station to India's rail network.

▷ **1883** One of the world's most glamorous trains is launched. From 1891 it is known as the *Orient Express*.

▷ **1885** The Canadian Pacific Railway's transcontinental route is completed.

▷ **1888** Bombay's Victoria Terminus is completed a decade after work started.

△ **Victorian Gothic**
Victoria Terminus, designed by the consulting British architect Frederick William Stevens, bears some resemblance to St Pancras railway station in London.

▷ **1888** British railway companies compete in the London to Edinburgh "Race to the North".

▷ **1890** The Forth Bridge opens, seven years after construction started.

▷ **1891** Work begins on one of the most ambitious engineering projects ever – the Trans-Siberian Railway.

19th-century Racers

The development of sleek express steam engines in the late 19th century led to publicity-seeking railways in both the US and UK competing for the fastest journey times on rival intercity routes. In the UK the famous "Races to the North" of 1888 and 1895 saw the railways of the rival East Coast and West Coast Main Lines between London and Scotland engage in a dangerous high-speed struggle for supremacy. In the US there was fierce competition between the Pennsylvania Railroad and the New York Central & Hudson River Railroad on their New York to Buffalo routes during the 1890s. This triggered electrifying performances by the latter company's celebrity locomotive No. 999 while hauling the *Empire State Express*.

◁ **GNR Stirling Single Class, 1870**

Wheel arrangement	4-2-2
Cylinders	2
Boiler pressure	170 psi (11.95 kg/sq cm)
Driving wheel diameter	97 in (2,464 mm)
Top speed	85 mph (137 km/h)

Patrick Stirling designed this locomotive for the Great Northern Railway. A total of 53 of these single-wheeler locomotives were built at Doncaster Works between 1870 and 1895. The locomotives hauled express trains on the East Coast Main Line between London King's Cross and York and were involved in the "Races to the North" of 1888 and 1895. No. 1, shown here, is preserved at the National Railway Museum in York, UK.

Races to the North

Headlined in newspapers as the "Race to the North", railway companies unofficially raced each other on two main lines between London and Edinburgh in 1888. The West Coast Main Line trains were operated by the London & North Western Railway and the Caledonian Railway; and the East Coast Main Line trains, by the Great Northern Railway, the North Eastern Railway, and the North British Railway. Following the completion of the s Forth Bridge in 1890, the companies raced between London and Aberdeen. After a derailment at Preston in 1896, the practice was banned and speed limits were enforced.

Record run A Caledonian Railway postcard shows Engine No. 17 and driver John Souter at Aberdeen after their race-winning run on 23 August 1895.

▷ **LNWR Improved Precedent Class, 1887**

Wheel arrangement	2-4-0
Cylinders	2 (inside)
Boiler pressure	150 psi (10.54 kg/sq cm)
Driving wheel diameter	81 in (2,057 mm)
Top speed	approx. 80 mph (129 km/h)

A total of 166 Improved Precedent Class express locomotives, designed by F.W. Webb, were built at the London & North Western Railway's Crewe Works between 1887 and 1901. No. 790 *Hardwicke* set a new speed record between Crewe and Carlisle during the "Race to the North" on 22 August 1895. It is preserved at the National Railway Museum in York, UK.

◁ CR No. 123, 1886

Wheel arrangement 4-2-2

Cylinders 2 (inside)

Boiler pressure 160 psi (11.25 kg/sq cm)

Driving wheel diameter 84 in (2,134 mm)

Top speed approx. 80 mph (129 km/h)

Built as an exhibition locomotive by Neilson & Co. of Glasgow for the Caledonian Railway in 1886, this unique single-wheeler hauled expresses between Carlisle and Glasgow. Following retirement in 1935 it was preserved and is now on display at the Riverside Museum in Glasgow.

▷ NYC&HR No. 999, 1893

Wheel arrangement 4-4-0

Cylinders 2

Boiler pressure 180 psi (12.65 kg/sq cm)

Driving wheel diameter 86½ in (2,197 mm)

Top speed approx. 86 mph (138 km/h)

Alleged to have travelled at over 100 mph (161 km/h), No. 999 was built in 1893 to haul the New York Central & Hudson River Railroad's flagship train, the *Empire State Express*, between New York and Buffalo. This celebrity locomotive was exhibited at the Chicago World's Fair before being retired in 1952. Nicknamed the "Queen of Speed", No. 999 is on display at the Chicago Museum of Science & Industry.

TECHNOLOGY

Standard Rail Time

Confusion reigned on the early railways as clocks at stations were set at local time, causing difficulty for railway staff and passengers alike. In the UK the Great Western Railway introduced a standardized "London Time" for their station schedules in 1840. This synchronization used Greenwich Mean Time (GMT) set by the Royal Observatory at Greenwich, which later became accepted as the global standard time. In 1883 railways in the US and Canada split both countries longitudinally into geographic time zones and introduced Railroad Standard Time.

Time regulation Made by American jeweller Webb C. Ball in 1889, this precision regulator clock helped maintain the accuracy of other timepieces on the Baltimore & Ohio Railroad.

▽ LB&SCR B1 Class, 1882

Wheel arrangement 0-4-2

Cylinders 2 (inside)

Boiler pressure 150 psi (10.53 kg/sq cm)

Driving wheel diameter 78 in (1,980 mm)

Top speed approx. 70 mph (113 km/h)

The B1 Class locomotives were designed by William Stroudley for the London, Brighton & South Coast Railway. A total of 36 were built at Brighton Works between 1882 and 1891. Hauling heavy expresses between London and Brighton, they were named after politicians, railway officials, or places served by the railway. The last survivor was retired in 1933, and No. 214 *Gladstone* is preserved at the National Railway Museum in York.

London Locals

Growing prosperity and personal mobility enabled people to move away from the centre of London. Railroads supplied transportation links from the new suburbs to the city, giving birth to the commuter train. While the Great Eastern Railway among others provided a peak-time, steam-hauled service, electric traction – overground and underground – was the future. The first deep-level "tube" line, the City & South London Railway, which opened in 1890 was the nucleus of London's underground system. Other cities soon followed London's example: Liverpool in northwest England and Budapest and Paris in Continental Europe. In the US, the Boston subway opened in 1897 and, by 1904, had been joined by New York's.

△ GWR 633 Class, 1871

Wheel arrangement	0-6-0T
Cylinders	2
Boiler pressure	165 psi (11.6 kg/sq cm)
Driving wheel diameter	54½ in (1,384 mm)
Top speed	approx. 40 mph (64 km/h)

Designed by George Armstrong and built at Wolverhampton Works, several of the 12-strong 633 Class were fitted with condensing apparatus to take Great Western Railway trains through the tunnels, so gaining the nickname "Tunnel Motors". Much modified, some lasted until 1934.

△ LB&SCR A1 Class, 1872

Wheel arrangement	0-6-0T
Cylinders	2
Boiler pressure	150 psi (10.53 kg/sq cm)
Driving wheel diameter	48 in (1,220 mm)
Top speed	approx. 60 mph (96 km/h)

The London, Brighton & South Coast Railway's suburban network was the domain of William Stroudley's small, six-coupled tanks. Fifty were built between 1872 and 1880, and the bark of their exhaust earned them the nickname of "Terriers". They were named after places they served, in the case of No. 54 *Waddon* (1875), a district near Croydon.

◁ NLR 75 Class, 1879

Wheel arrangement	0-6-0T
Cylinders	2
Boiler pressure	160 psi (11.24 kg/sq cm)
Driving wheel diameter	52 in (1,321 mm)
Top speed	approx. 30 mph (48 km/h)

John C. Park supplied the North London Railway with this shunting engine to serve the dock system around Poplar. Thirty were built up to 1905 and, as they rarely left the docks, no coal bunker was fitted; fuel was stored on the footplate.

◁ LSWR 415 Class, 1882

Wheel arrangement	4-4-2T
Cylinders	2
Boiler pressure	160 psi (11.25 kg/sq cm)
Driving wheel diameter	67 in (1,702 mm)
Top speed	approx. 45 mph (72 km/h)

Designed by William Adams of the London & South Western Railway, 71 of the 415 Class were built from 1882 to 1885. Put to work on suburban services out of London's Waterloo, three ended their days on the southwest Lyme Regis branch, where their short wheelbase and leading bogie were ideal to negotiate the severe curves.

London's Carriages

Both the Metropolitan and District railways began by using locomotive-hauled carriages. However, few offered the upholstered luxury of the Metropolitan's "Jubilee" coach. Most followed the pattern of the District's No. 100, with 10 passengers to each compartment. The distinction between the classes even extended to lighting: first class travellers enjoyed two gas jets, while second and third class passengers made do with one. Conditions improved little with the coming of the City & South London Railway, which became known as the "sardine tin railway".

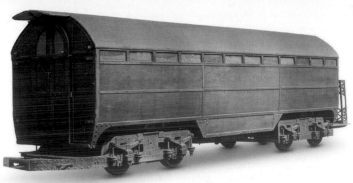

◁ C&SLR "Padded Cell", 1890

Type	underground passenger carriage
Capacity	32 passengers
Construction	wooden body on two 4-wheel bogies
Railway	City & South London Railway

Tunnel diameter restricted carriage size on this first "tube" line. Coaches were fitted with high-backed seating, running along the length, and gates at either end to allow passengers on and off. With the only windows being slits above seats, and air entering through roof ventilators, the nickname "padded cells" was appropriate.

△ **C&SLR electric locomotive, 1889**

Wheel arrangement	0-4-0 (Bo)
Power supply	0.5 kV DC third rail
Power rating	100 hp (74.60 kW)
Top speed	25 mph (40 km/h)

The first important railway to use electric traction was the City & South London Railway. When opened in 1890 the line had six stations and ran from City to Stockwell. Operated by 14 locomotives, one of which – with a train of later steel-bodied carriages with full-length windows – is passing Borough Junction in this 1922 photograph.

TALKING POINT
Cemetery Railways

As London's population doubled in the 19th century, burying the dead became a crisis. The boldest solution came from Sir Richard Broun and Richard Sprye – their scheme involved buying a large piece of land away from the city but with a direct rail link to London. The chosen location was Brookwood, Surrey, 23 miles (37 km) along the LSWR main line out of Waterloo. They envisaged that coffins would be brought to Brookwood either late at night or early in the morning with mourners travelling by dedicated train services during the day.

Burying the dead The London Necropolis (Greek for "city of the dead") Railway opened in 1854. After the terminus was bombed in 1941, its services never ran again.

▽ **GER S56 Class, 1886**

Wheel arrangement	0-6-0T
Cylinders	2
Boiler pressure	180 psi (12.65 kg/sq cm)
Driving wheel diameter	48 in (1,220 mm)
Top speed	approx. 60 mph (96 km/h)

James Holden designed a small but powerful, six-coupled tank for the Great Eastern Railway's inner-suburban services in 1886. It was equipped with Westinghouse compressed air brakes, ideal where stations were close together. The sole survivor, No. 87 of 1904, is part of Britain's National Collection.

△ **Met C Class, 1891**

Wheel arrangement	0-4-4T
Cylinders	2
Boiler pressure	140 psi (9.84 kg/sq cm)
Driving wheel diameter	66 in (1,676 mm)
Top speed	approx. 60 mph (96 km/h)

The Metropolitan Railway's C Class consisted of just four engines built by Neilson & Co. of Glasgow. After the Met's expansion into Hertfordshire and Buckinghamshire, they hauled trains from the city out to Watford, Amersham, and Aylesbury.

▷ **Met Jubilee Coach No. 353, 1892**

Type	four-compartment, first-class passenger coach
Capacity	32
Construction	original wooden body on later 4-wheel steel chassis
Railway	Metropolitan Railway

This carriage served the Metropolitan Railway from 1892 until 1907 when it was sold to the Weston, Clevedon & Portishead Light Railway. Restored to mark the railway's 150th anniversary in 2013, it is now at the London Transport Museum.

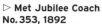

△ **DR Coach No. 100, 1884**

Type	four-compartment, third-class passenger carriage
Capacity	40
Construction	original wooden body on later 4-wheel steel chassis
Railway	District Railway

The origins of coach No. 100 of the District Railway are uncertain. What is definite is that the body finished up as a storage shed in Kent. It was rescued, placed on a new chassis, and now runs on the Kent & East Sussex Railway, where a District Railway brown livery was applied.

End of the Great Western Broad Gauge

When the broad-gauge rails of the Great Western Railway (GWR) first came up against those of a narrower-gauge company at Gloucester, England in 1844, passengers were forced to change trains to continue their journey to the north or the southwest. The impracticality of mismatched gauges led the UK government to set up a Gauge Committee to examine the issue. While the committee agreed that the GWR offered greater speed and stability (due as much to the excellence of Isambard Kingdom Brunel's railway and Daniel Gooch's locomotives as the width of the track), it concluded that the narrower gauge suited its long-term interest. A change to the narrower gauge (now known as standard gauge) became inevitable.

Firstly, a third rail to take standard-gauge trains was laid on the GWR, reaching Paddington in 1861. Gauge conversion began in 1866 and took almost three decades to complete. The final change, along the West of England main line, took place over a weekend in May 1892. It was meticulously planned with 4,200 workmen positioned along the line and prefabricated track sections such as facing points and crossovers were carried to where they were needed. On 23 May 1892, the operation was completed and Brunel's broad gauge was consigned to history.

On 20 May 1892, the final *Cornishman* broad-gauge express left London, Paddington for Penzance with Rover Class 4-2-2 *Great Western* at its head.

C&PA Snow Plow

The Coudersport & Port Allegany Railroad (C&PA) Snow Plow is typical of the wooden ploughs built by the Russell Company of Ridgeway, Pennsylvania, from the late 19th century onwards. Designed to be pushed by one or two steam engines along a single-track line, it was fitted with a flange that scraped snow and ice from the insides of the rails, creating a groove for the flanges of engine and carriage wheels.

IN WINTER, HEAVY SNOWFALLS and icy conditions regularly closed railways in the US and Canada. Faced with a loss of business, the railway companies started to use wooden wedges attached to the front of locomotives to clear snow from the tracks. By the late 19th century this makeshift arrangement was superceded by the introduction of separate snow-plough wagons mounted on bogies and pushed by locomotives.

Believed to be oldest snow plough of its type in existence, the C&PA Snow Plow is a wedge-style model that was built around 1890 under licence for the Russell Snow Plow Company by the Ensign Manufacturing Company in Huntingdon, West Virginia. A cupola was fitted on the roof directly behind the plough blades to give the crew a view of the track ahead. The plough was used on the C&PA until 1945, when it was damaged in an accident. It later became the property of the Wellsville, Addison & Galeton Railroad before being donated to the Railroad Museum of Pennsylvania in 1980, where it has since been restored.

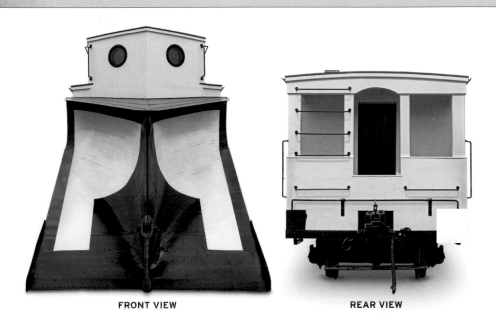

FRONT VIEW

REAR VIEW

Short-line railway
Opened in 1882, the Coudersport & Port Allegany Railroad (C&PA) was a 32-mile- (50-km-) timber-carrying short line in Potter and McKean Counties in Pennsylvania. It was abandoned in the early 1970s.

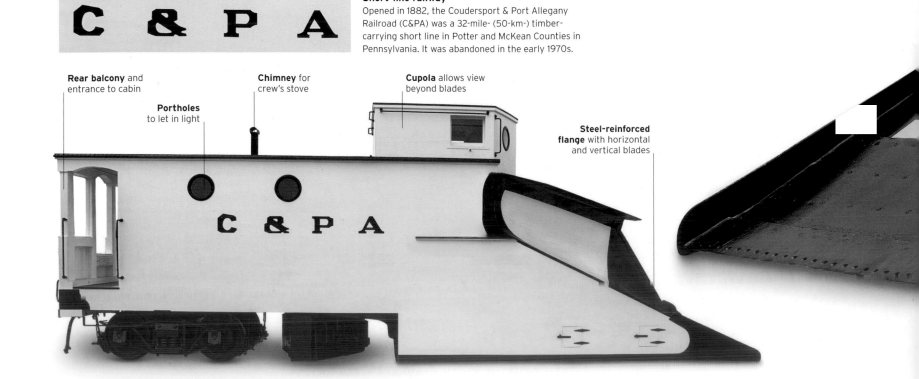

Rear balcony and entrance to cabin

Portholes to let in light

Chimney for crew's stove

Cupola allows view beyond blades

Steel-reinforced flange with horizontal and vertical blades

SPECIFICATIONS	
Type	Wedge-type snow plough
Origin	USA
Designer/builder	Russell Co.
Number produced	1
In-service period	c.1890-1945
Weight	Not known
Construction	Wood and steel
Railway	Coudersport & Port Allegany Railroad

Blades at work

As the plough was pushed from behind by one or two steam locomotives, the sharp front blade would lift snow off the track before deflecting it to either side with the angled vertical blade.

EXTERIOR

While the main body of the C&PA Snow Plow was constructed of seasoned hardwood, the impressive plough blades were reinforced with steel. The plough was mounted on two four-wheeled bogies, one of which was concealed beneath the front blade housing.

1. Porthole-style windows above top edge of plough **2.** Front coupling bar **3.** Rivets on front edge of plough **4.** Chimney **5.** Round and square windows looking out from the observation level **6.** Journal box access door **7.** Journal box, which contains the journal bearing **8.** Flange (secondary plough) **9.** Back wheel brake shoe **10.** Bogie at rear (arch bar truck) **11.** Deck behind cabin **12.** Bars across aperture on platform, used as a ladder to access roof **13.** Coupling ring at rear **14.** Angle cock **15.** Coupling at rear

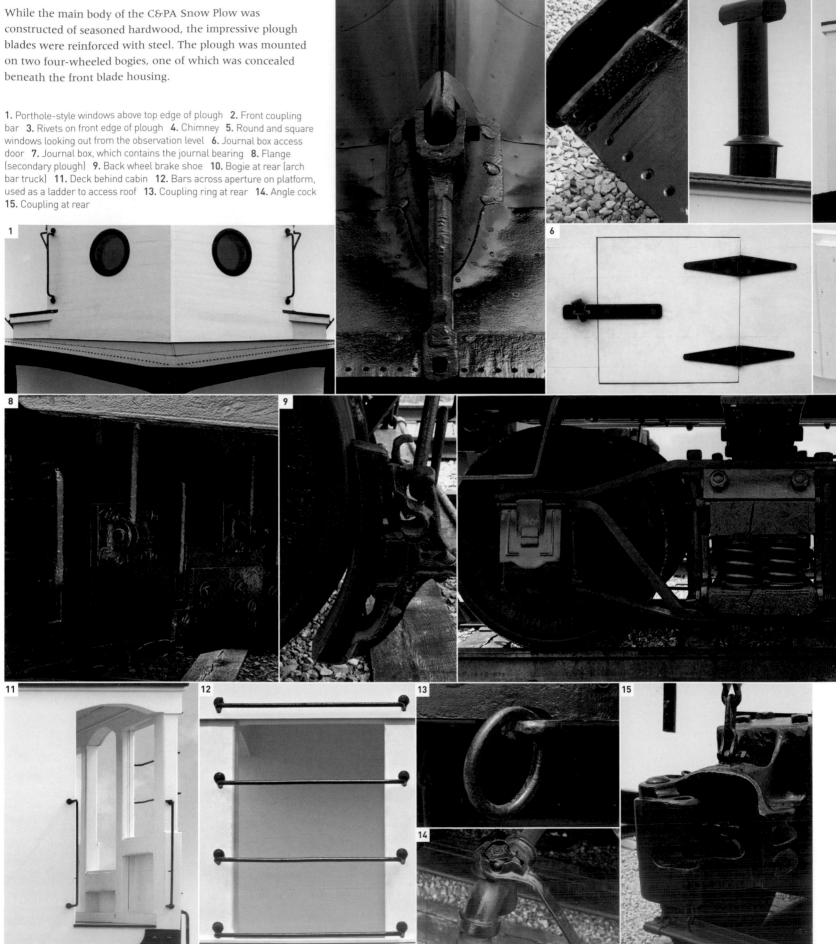

INTERIOR

The cabin was reinforced at the front end by steel girders to prevent the crew being crushed during snow-clearing operations. It was fitted out with a handbrake, air pressure gauge, steps for the cupola forward lookout, and a coal-fired safety stove fitted with a flange on top to prevent pans from sliding off. The suspended flange beneath the cabin floor could be raised or lowered either manually or by air pressure, to avoid damaging pointwork (switches) and level crossings.

16. Cabin interior   17. Suspension springs to the flange   18. Air reservoir pipes behind steps   19. Piston to adjust height of flange   20. Brake wheel   21. Base of brake wheel with cog mechanism   22. Air brake pressure gauge   23. Coal stove   24. Pennsylvania Railroad stamp on stove   25. Decorative door handle

Delivering to America

The South Carolina Railroad began conveying mail as early as 1833, but the first regular mail service in the US did not start until two years later, on the Baltimore and Ohio. In July 1838, US Congress approved the use of all rail routes to carry mail, earning the railroad companies significant revenues from the US Postal Department (USPOD).

In 1862 US President Abraham Lincoln approved the building of a 1,928-mile (3,084-km) line between Omaha, Nebraska, and Sacramento, California, to bring people, trade, and a vitally needed postal service to the western regions. That same year the government unified mail traffic under the Railway Mail Service, which led to the construction of dedicated Railway Post Office

(RPO) cars. The first permanent RPO service ran between Chicago, Illinois, and Clinton, Iowa. By 1869 the transcontinental railroad was complete, giving postal cars the means to carry mail across the breadth of the US. At their peak, RPO cars covered more than 200,000 miles (320,000 km) on more than 9,000 routes.

The US also ran the world's first postal express, which left New York's Grand Central Station for Chicago on 16 September 1875. It completed the journey in just over 24 hours, and became the forerunner for night mail trains across the world.

The rail junction at night depicted in this Currier & Ives print of 1876 shows some of the express train types that carried mail to major US cities.

Building Great Railways
Canadian Pacific

Opened across the vast spaces of the Prairies and through the Rocky Mountains in 1886, the Canadian Pacific Railway was Canada's first transcontinental line, linking Vancouver on the Pacific west coast with Montreal on the St Lawrence River.

IN 1871 THE GOVERNMENT of the recently formed Dominion of Canada promised the isolated western province of British Columbia that a railway would be built across the Rocky Mountains within 10 years. The project got off to a slow start and by 1880 only 300 miles (483 km) of line had been built.

However, in 1881 a group of Canadian businessmen formed the Canadian Pacific Railway (CP) and, with financial assistance and land from the government, took over the unfinished lines and recommenced construction work from both the east and west. Overseen by the new general manager of the railway, William Cornelius Van Horne, tracklaying in the east began at Bonfield, north of the Great Lakes, and proceeded slowly westwards across the remote, sparsely populated

Colonizing Canada
Canadian Pacific offered packages of sea and rail travel to immigrants.

and lake-strewn landscape of the Canadian Shield in Ontario towards Winnipeg. The link connecting the new railway at Bonfield with the eastern cities of Ottawa and Montreal had already been built by the Canadian Central Railway and the Ontario & Quebec Railway, both of which the CP leased from 1884.

From Winnipeg, construction continued westwards across the vast plains of Saskatchewan to Calgary at the foot of the Rockies. From Calgary, gangs of Chinese labourers built the railway up into the Rockies through Banff, reaching Kicking Horse Pass in 1884. From here the railway made a steep descent down the Big Hill before climbing again to cross the Selkirk Range at Rogers Pass.

Climbing the Rocky Mountains
A passenger train with vista dome cars and an observation car follows the Bow River on the CP's scenic route through the Banff National Park in the Canadian Rockies.

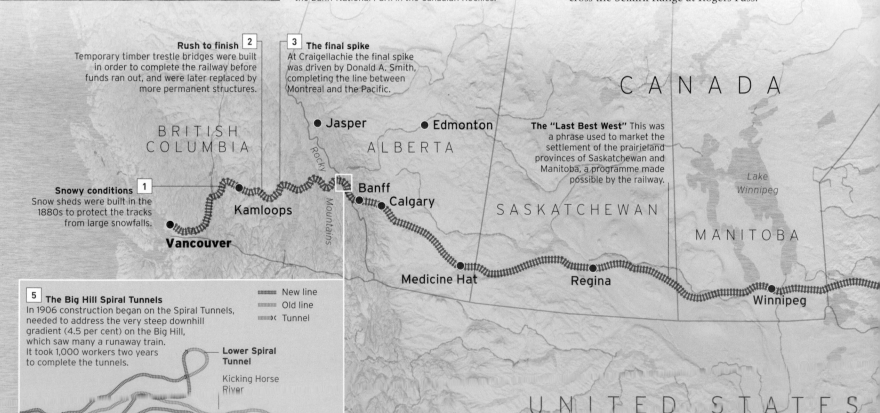

2 Rush to finish
Temporary timber trestle bridges were built in order to complete the railway before funds ran out, and were later replaced by more permanent structures.

3 The final spike
At Craigellachie the final spike was driven by Donald A. Smith, completing the line between Montreal and the Pacific.

The "Last Best West" This was a phrase used to market the settlement of the prairieland provinces of Saskatchewan and Manitoba, a programme made possible by the railway.

1 Snowy conditions
Snow sheds were built in the 1880s to protect the tracks from large snowfalls.

BRITISH COLUMBIA

Kamloops

Vancouver

Jasper

Rocky Mountains

ALBERTA

Edmonton

Banff
Calgary

Medicine Hat

SASKATCHEWAN

Regina

CANADA

Lake Winnipeg

MANITOBA

Winnipeg

5 The Big Hill Spiral Tunnels
In 1906 construction began on the Spiral Tunnels, needed to address the very steep downhill gradient (4.5 per cent) on the Big Hill, which saw many a runaway train. It took 1,000 workers two years to complete the tunnels.

— New line
— Old line
—)(Tunnel

Lower Spiral Tunnel

Kicking Horse River

Upper Spiral Tunnel

UNITED STATES OF AMERICA

Across the Canadian Prairies
A train travels over the prairie near Morse, between Regina and Medicine Hat, in Saskatchewan. Natural gas was discovered in the prairies by workers constructing the line.

To the west of the Rockies, construction continued through the Monashee Mountains before the two lines met at Craigellachie, where a ceremonial final spike was driven in 1885. The entire route was now complete and the first transcontinental train ran between Montreal in the east and Port Moody in the west in 1886. A year later the western terminus was moved to Vancouver. Attracted by a CP package deal, which included passage on a company ship, travel on a company train, and land sold by CP, thousands of immigrants from Europe were soon streaming westwards on the new railway in search of new lives.

The Big Hill, with its treacherously steep gradients, was bypassed in 1909 when a series of Spiral Tunnels were opened, and the steep gradient up to Rogers Pass was also later bypassed by the opening of the Connaught Tunnel in 1916.

KEY FACTS

DATES
1881 Construction begins at Bonfield
1882 Thunder Bay branch completed
1885 3 November: final spike on Lake Superior section; 7 November: Final spike at Craigellachie, BC
1886 28 June: First transcontinental passenger service leaves Dalhousie Station, Montreal
1909 Spiral Tunnels at Kicking Horse pass open
1978 CP passenger services taken over by Via Rail
1990 *The Canadian* passenger train rerouted over Canadian National Railways route

FIRST PASSENGER TRAIN
Locomotive type American Standard 4-4-0 steam
Carriages 2 baggage cars, 1 mail car, 1 second-class coach, 2 immigrant sleepers, 2 first-class coaches, 2 sleeping cars, and a diner

JOURNEY
Montreal to Port Moody (1886) 2,883 miles (4,640 km); 6 days, 6 nights
Montreal to Vancouver (1963) 2,888 miles (4.648 km); 69 hours

RAILWAY
Gauge Standard 4 ft 8 ½ in (1.434 m)
Tunnels Connaught Tunnel 5 miles (8 km); Spiral Tunnel No. 1: 3,153 ft (961 m), Tunnel No. 2: 2,844 ft (867 m)
Bridges Stoney Creek Bridge 300 ft (91 m) high
Highest point 5,338 ft (1,627 m) Kicking Horse Pass

KEY
● Start/Finish
● Main stations
⊞⊞⊞ Main route

Muskeg terrain This required sections of the track to be elevated to prevent it sinking during thaws.

Bonfield This was the site of the first track to be laid.

ONTARIO

QUÉBEC

4 First passenger train
The first transcontinental service left Dalhousie Station, Montreal on 28 June 1886.

Thunder Bay

Lake Superior

Lake Michigan

Lake Huron

Lake Ontario

Montreal

Ottawa

PEAKS AND VALLEYS

Construction in the Rocky Mountains was particularly perilous. Workers faced harsh terrain, the threat of forest fires, heavy snowfall, and avalanches as they built the line across deep valleys, up steep gradients, and through rock.

Specialist Steam

Initially used for hauling coal, railways were soon adapted to play similar roles in the fast-growing industrial landscape. Narrow-gauge lines and engines were ideal for quarries, foundries, shipyards, brickworks, and some military sites. Dock railways required small but powerful engines that could weave their way along quaysides, while in chemical plants and munitions factories the danger posed by stray sparks was overcome by developing fireless locomotives. Ingenious engines and track were used to scale mountains. There were few places where the steam locomotive could not serve.

◁ **VRB No. 7, 1873**

Wheel arrangement	0-4-0VBT
Cylinders	2
Boiler pressure	185 psi (13 kg/sq cm)
Driving wheel diameter	25 in (644 mm)
Top speed	approx. 5 mph (8 km/h)

Designed by Niklaus Riggenbach and built by the Swiss Locomotive Co., No. 7 was employed on the Vitznau-Rigi mountain railway (Vitznau-Rigi Bahn, or VRB) near Lucerne, Switzerland, until 1937. Its vertical boiler kept a safe water level on the steep climb, which was undertaken using a rack-and-pinion system.

▷ **SRR A-4 Class "Camelback", 1877**

Wheel arrangement	0-4-0
Cylinders	2
Boiler pressure	200 psi (14.06 kg/sq cm)
Driving wheel diameter	50 in (1,270 mm)
Top speed	approx. 20 mph (32 km/h)

Engines of Pennsylvania's coal-carrying railways were fired on cheap anthracite waste that needed a large firebox for ample combustion, so the driver's cab could not be sited behind it. Instead, it straddled the firebox, hence the nickname "Camelback". No. 4 worked on the Philadelphia & Reading Railroad and the Strasburg Railroad.

◁ **FR Double Fairlie No. 10** *Merddin Emrys*, 1879

Wheel arrangement	0-4-4-0T
Cylinders	4
Boiler pressure	160 psi (11.25 kg/sq cm)
Driving wheel diameter	32 in (813 mm)
Top speed	approx. 35 mph (56 km/h)

Following a design by British engineer Robert Fairlie, *Merddin Emrys* was the first locomotive built by the Ffestiniog Railway's workshops. A double-ended, articulated tank engine riding on powered bogies, today's No. 10 is much rebuilt.

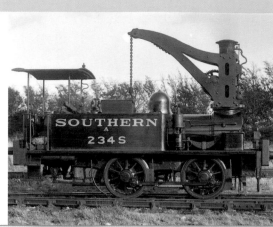

Southern Railway No. 234S, 1881 This crane tank was used at Ashford Locomotive Works and Folkestone Harbour, both in Kent, and at Lancing Carriage Works in Sussex. It was retired in 1949.

△ LYR *Wren*, 1887

Wheel arrangement	0-4-0ST
Cylinders	2
Boiler pressure	170 psi (11.95 kg/sq cm)
Driving wheel diameter	16½ in (418 mm)
Top speed	approx. 5 mph (8 km/h)

Wren was one of eight small saddletanks employed on the 7½-mile- (12-km-), 1-ft 6-in- (0.46-m-) gauge track serving the Lancashire & Yorkshire Railway's works at Horwich, Lancashire. The engine was built by Beyer Peacock & Co. of Manchester, and remained in use until 1962.

◁ Hunslet *Lilla*, 1891

Wheel arrangement	0-4-0ST
Cylinders	2
Boiler pressure	120 psi (8.43 kg/sq cm)
Driving wheel diameter	28 in (660 mm)
Top speed	approx. 10-12 mph (16-19 km/h)

Lilla is a survivor from 50 saddletanks built by the Hunslet Engine Co. of Leeds, England, between 1870 and 1932 for Welsh slate quarries. It was retired from Penrhyn Quarry in 1957 and is now preserved on the Ffestiniog Railway in North Wales.

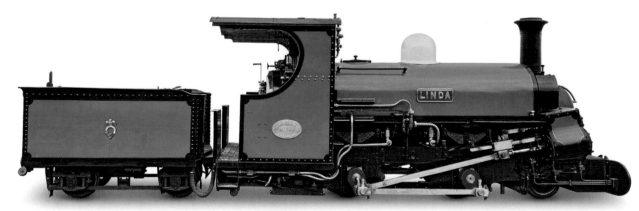

◁ Hunslet *Linda*, 1893

Wheel arrangement	0-4-0STT
Cylinders	2
Boiler pressure	140 psi (9.9 kg/sq cm)
Driving wheel diameter	26 in (660 mm)
Top speed	approx. 12-18 mph (19-29 km/h)

From the same stable as *Lilla* but more powerful, *Linda* was used on the Penrhyn Quarry's "mainline", which ran from Bethesda to Port Penrhyn, near Bangor, Wales. Another Ffestiniog veteran, *Linda* has been rebuilt there as a 2-4-0 saddletank tender engine.

▷ Saxon IV K Class, 1892

Wheel arrangement	0-4-4-0T
Cylinders	4 (compound)
Boiler pressure	174 psi/203 psi/217 psi (12.23 kg/sq cm/14.27 kg/sq cm/ 15.25 kg/sq cm) (variations within class)
Driving wheel diameter	30 in (760 mm)
Top speed	approx. 19 mph (30 km/h)

Germany's most numerous narrow-gauge class, 96 of these were built for the Royal Saxon State Railways from 1892 to 1921. They were articulated, and used the Günther-Meyer system of powered bogies; only 22 survive.

Merddin Emrys

The FR Double Fairlie No.10 *Merddin Emrys* was built to combine large haulage capacity with route flexibility. Originally designed by Robert Francis Fairlie and championed by the Ffestiniog Railway in North Wales, Double Fairlie articulated locomotives were able to negotiate tight curves thanks to their flexible steam pipes and pivoting power bogies. Fairlie's patented design was also used in Russia, Mexico, Germany, Canada, Australia, and the US.

ON 21 JULY 1879, almost 10 years since the first of Robert Fairlie's double-ended articulated locomotives had arrived on the Ffestiniog Railway (FR), *Merddin Emrys* was rolled out of the railway's Boston Lodge workshops. Designed by G.P. Spooner using Fairlie's principles, No.10 *Merddin Emrys* was the third Double Fairlie to be employed on the FR and it can still be seen there today. The locomotive could comfortably haul 80-ton (81-tonne) loads uphill, from Porthmadog to the slate quarries at Blaenau Ffestiniog 13 miles (21 km) away. Impressively, some of these trains were up to 1,312 ft (400 m) long.

The design featured a double-ended boiler with two separate fireboxes in the centre. Unlike conventional steam locomotives that carried their boilers on a rigid frame, the boiler and superstructure of the Double Fairlie were supported at each end by a short-wheelbase power bogie, connected by flexible steam hoses. This allowed the bogies to turn into a curve before the main body of the locomotive. It was possible to drive each "end" of the locomotive independently, with the driver and fireman standing on either side of the firebox.

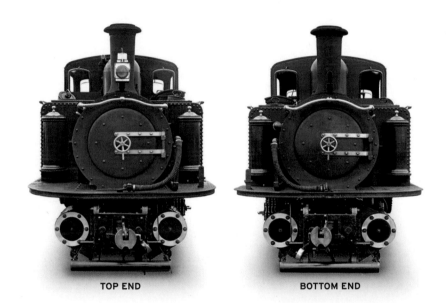

TOP END   BOTTOM END

Six of the best
Built in 1836, the Ffestiniog Railway used six "Double Fairlie" 0-4-4-0T locomotives to transport slate from Blaenau Ffestiniog to the sea at Porthmadog, South Wales.

SPECIFICATIONS			
Class	FR Double Fairlie	**In-service period**	1879-present (Merddin Emrys)
Wheel arrangement	0-4-4-0T	**Cylinders**	4
Origin	UK	**Boiler pressure**	160 psi (11.25 kg/sq cm)
Designer/Builder	R. Fairlie/G.P. Spooner/FR	**Driving wheel diameter**	32 in (812 mm)
Number produced	6 (2 of this improved design)	**Top speed**	35 mph (56 km/h)

Sandbox in front of side tank

Steam dome on each boiler barrel

Driver's cab is split in half by boiler

Coal bunker within water tank

Separate exhausts serve each end

Side water tanks with 667-gallon (3,032-litre) capacity

Four-coupled power bogie on each end

Twin role
The fireman of a Double Fairlie has to contend with twice the amount of work. There are two fire boxes, but only one boiler, and a common water space. Both fireboxes need to be used to maintain working boiler pressure.

EXTERIOR

While it might appear to be a product of Victorian times, today's *Merddin Emrys* is virtually a new locomotive. In 1970 it was extensively rebuilt with a new boiler, which gave it a larger, less traditional look. By 1973 it was converted to burn oil instead of coal, and by 1984 it was in need of another overhaul. Its builders decided to remake *Merddin Emrys* in its original 1879 appearance, but retained its larger superstructure in line with the Ffestiniog Railway's improved loading gauge restrictions. The "new" locomotive emerged in 1988, only to be overhauled again in 2005. *Merddin Emrys* reverted to being a coal burner in 2007.

1. Nameplate 2. Smokebox door 3. Water tank filler 4. Number plate on smokebox 5. Sandbox 6. Top end whistle 7. Mechanical lubricator 8. Reverser lever attached to boiler 9. Crosshead 10. Handbrake attached to boiler 11. Bottom end coal bunker 12. Bottom end driver's side bogie 13. Crosshead and cylinder 14. Top of driving wheel 15. Speedometer drive 16. Small whistle 17. "Norwegian Chopper" coupler

CAB INTERIOR

The large fireboxes in the centre of the cab mean that the engine crew have to stand in confined spaces on either side of the footplate with the firebox between them. The driver has a single reverser and two regulator handles, which allow the necessary amount of steam to be thrust to either power bogie, as and when required. The design and position of the handles enable regulators to be opened simultaneously with one hand. The fireman, meanwhile, has two firehole doors, one for each firebox, and two sets of gauges. Coal is carried in bunkers built into the water tanks on the fireman's side.

18. Bottom end firebox **19.** Water gauge **20.** Boiler pressure gauge **21.** Top end manifold shutoff **22.** Coal bunker door **23.** Vacuum ejector, steam brake, and injector **24.** Vacuum release valve **25.** Top end injector and slacker valve **26.** Top end firebox door

Shrinking the World

The introduction of steam engines on the narrow-gauge, slate-carrying railway at Ffestiniog in Wales in 1863 had led to the adoption of other narrower-gauge railways around the world. These lines were suited to mountainous regions as they were cheaper to construct and could cope with sharper curves and steeper gradients. In the 1870s India built its first locomotive using parts imported from Britain, and in 1872 Japan opened its first railway. Elsewhere, larger engines were being introduced and the mass production of freight locomotives had begun.

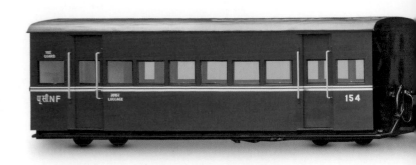

◁ **Japan's No. 1, 1871/2**

Wheel arrangement	2-4-0T
Cylinders	2
Boiler pressure	140 psi (10 kg/sq cm)
Driving wheel diameter	52 in (1,320 mm)
Top speed	approx. 30 mph (48 km/h)

Built in the UK by the Vulcan Foundry in 1871, No. 1 was the first steam locomotive to operate on Japan's inaugural public railway, from Tokyo to Yokohama, which opened in 1872. From 1880 it went to work on other Japanese railways before retiring in 1930. It is now on display at the Saitama Railway Museum.

▷ **V&TRR No. 20 _Tahoe_, 1875**

Wheel arrangement	2-6-0
Cylinders	2
Boiler pressure	130 psi (9.14 kg/sq cm)
Driving wheel diameter	48 in (1,220 mm)
Top speed	approx. 30 mph (48 km/h)

Built by the Baldwin Locomotive Works, Philadelphia, in 1875, No. 20 _Tahoe_ worked on the Virginia & Truckee Railroad in Nevada, US, until 1926. The 41.88-ton (38-tonne) locomotive was temporarily brought out of retirement during WWII. It has since been restored and is now on display at the Railroad Museum of Pennsylvania in Strasburg.

TALKING POINT

Prince of Wales's Coach

Constructed at the Agra Workshops of the 3-ft 3-in- (1-m-) gauge Rajputana Malwa Railway in 1875, this elegant coach was specially built for the then Prince of Wales (later King Edward VII) for his visit to India in 1877. The prince travelled to India for the Royal Durbar, which celebrated the coronation of his mother Queen Victoria as Empress of India. With all of its original fittings intact, this coach is now on display at the National Rail Museum, New Delhi.

Royal transport This unique, four-wheel coach features balconies at each end with seating for four armed guards. The carriage has sunshades on both sides and is decorated with emblems of the British Crown.

△ **Indian F Class, 1874**

Wheel arrangement	0-6-0
Cylinders	2
Boiler pressure	approx. 140 psi (10 kg/sq cm)
Driving wheel diameter	approx. 57 in (1,448 mm)
Top speed	approx. 30 mph (48 km/h)

Derived from the British-built 3ft 3-in- (1-m-) gauge F Class mixed traffic locomotives introduced in 1874, F1 Class No. 734 was the first locomotive to be assembled in India, using imported parts. It worked on the Rajputana Malwa Railway from 1895, and is now an exhibit at the National Rail Museum, New Delhi.

▷ FR Single Fairlie *Taliesin*, 1876

Wheel arrangement 0-4-4T

Cylinders 2

Boiler pressure 150 psi (10.53 kg/sq cm)

Driving wheel diameter 32 in (810 mm)

Top speed approx. 20 mph (32 km/h)

Built for the 1-ft 11½-in- (0.60-m-) gauge Ffestiniog Railway in North Wales by the Vulcan Foundry, Single Fairlie *Taliesin* worked slate and passenger trains between Blaenau Ffestiniog and Porthmadog until withdrawn and scrapped in 1935. A working replica, using a few parts from the original engine, was built at the railway's Boston Lodge Workshops in 1999.

△ DHR Class B, 1889

Wheel arrangement 0-4-0ST

Cylinders 2

Boiler pressure 140 psi (10 kg/sq cm)

Driving wheel diameter 26 in (660 mm)

Top speed approx. 20 mph (32 km/h)

A total of 34 of these locomotives were built by Sharp Stewart & Co. and others for the 2-ft- (0.60-m-) gauge Darjeeling Himalayan Railway in India from 1889 to 1927. Some of them still run on this steeply graded line, which was declared a World Heritage Site by UNESCO in 1999.

△ Russian O Class, 1890

Wheel arrangement 0-8-0

Cylinders 2

Boiler pressure 156-213 psi (11-15 kg/sq cm)

Driving wheel diameter 47¼ in (1,200 mm)

Top speed approx. 35 mph (56 km/h)

Over 9,000 of the Russian O Class freight engines were built between 1890 and 1928, making it the second most numerous class of steam locomotives in the world. Armoured versions of this class were widely used to haul trains during WWI, the Russian Civil War, and WWII.

▷ CGR Class 7, 1892

Wheel arrangement 4-8-0

Cylinders 2

Boiler pressure 160-180 psi (11.25-12.65 kg/sq cm)

Driving wheel diameter 42½ in (1,080 mm)

Top speed approx. 35 mph (56 km/h)

Thirty-eight of these powerful freight locomotives were built in Scotland in 1892 for the 3-ft 6-in- (1.06-m-) gauge Cape Government Railway in South Africa. They worked on the newly formed South African Railways from 1912, until their withdrawal in 1972. Some saw service on the Zambesi Sawmills Railway in Zambia.

DHR B Class No. 19

If any class of locomotive defines a railway line it is the Darjeeling Himalayan Railway B Class. For many years these small, yet powerful, locomotives have hauled trains on the adhesion-worked mountain railway that climbs from the plains of northwest India through tea plantations to the hill station of Darjeeling. The idyllic scenery of the route has inspired many poetic descriptions, including "halfway to heaven" and "railway to the clouds".

THE FIRST FOUR B Class for the Darjeeling Himalayan Railway (DHR) were built by UK-based Sharp, Stewart, & Company in 1889. By 1927 the North British Locomotive Company of Glasgow, the Baldwin Locomotive Works of Philadelphia in the US, and the railway's own Tindharia Works had built a further 25. An additional five had been built for the Raipur Forest Tramway in 1925. After decades of service, four from the DHR stock were transferred to the Tipong Colliery Railway in 1970. Nowadays, some B Class still run on the DHR, while several exist as retired exhibits around India, and one was transferred to operate on the Matheran Hill Railway in 2002.

B Class No. 19 was sold to an American DHR enthusiast in 1962. After several years out of service the engine was bought by a British enthusiast, who restored it for use on the private Beeches Light Railway in Oxfordshire, UK.

FRONT VIEW OF ENGINE

REAR VIEW OF BRAKE CARRIAGE

SPECIFICATIONS	
Class	B
Wheel arrangement	0-4-0ST
Origin	UK
Designer/builder	Sharp, Stewart & Co.
Number produced	34 B Class
In-service period	1889 to date (No. 19)
Cylinders	2
Boiler pressure	140 psi (10 kg/sq cm)
Driving wheel diameter	26 in (660 mm)
Top speed	approx. 20 mph (32 km/h)

N.F.

Going up
Managed by India's Northeast Frontier Railway (NF), the DHR is 48 miles (78 km) long; it climbs from 328 ft (100 m) above sea level at New Jalpaiguri to 7,218 ft (2,200 m) at Darjeeling. The DHR is a UNESCO World Heritage Site.

Tender to carry air brake compressor and coal (not used on B Class in service on the DHR)

Cab has been raised to accommodate taller people

Saddle tank has 120-gallon (545-litre) water capacity

Coal bunker has 1,500-lb (680-kg) capacity

Original boiler
Although the B Class No.19 has been
overhauled for use in the UK, it retains
its original boiler dating back to 1889.
This is a remarkable feature that
is found in very few locomotives of
this vintage.

LOCOMOTIVE EXTERIOR

The short wheelbase of the B Class is ideally suited to the DHR's many curves and puts all of the locomotive's weight onto the rails for adhesion. DHR trains normally have a crew of nine: the driver, engineer, and fireman, a coal breaker who travels on the coal bunker in front of the cab, two sanders ride on the front to sand wet rails, and a guard and brakeman for each coach.

1. Engine number in English and Hindi **2.** Headlight and chimney **3.** Decoration on smokebox door securing dart **4.** "Chopper" coupling in style of Festiniog Railway **5.** Drain cock to enable water to be drained out of smokebox **6.** Brass lubrication box for steam glands **7.** Filler hatch for water tanks **8.** Safety valves **9.** Front of steam cylinder **10.** Cylinder block with steam cylinder below, valve above **11.** Original sand box **12.** Turbo alternator for head and cab lights **13.** Clack (non-return) valve and brass oil reservoir for axleboxes **14.** Isolating valve, on side of dome, for steam supply to driver's vacuum brake valve **15.** Mechanical lubricator for cylinders **16.** Left leading axle showing crosshead **17.** Right trailing bearer spring **18.** Left trailing coupling and connecting rod bearings **19.** Modern sand box on top of engine **20.** Top of engine showing empty former coal bunker **21.** Steam "fountain" and whistle in front of cab **22.** Handrail on tender

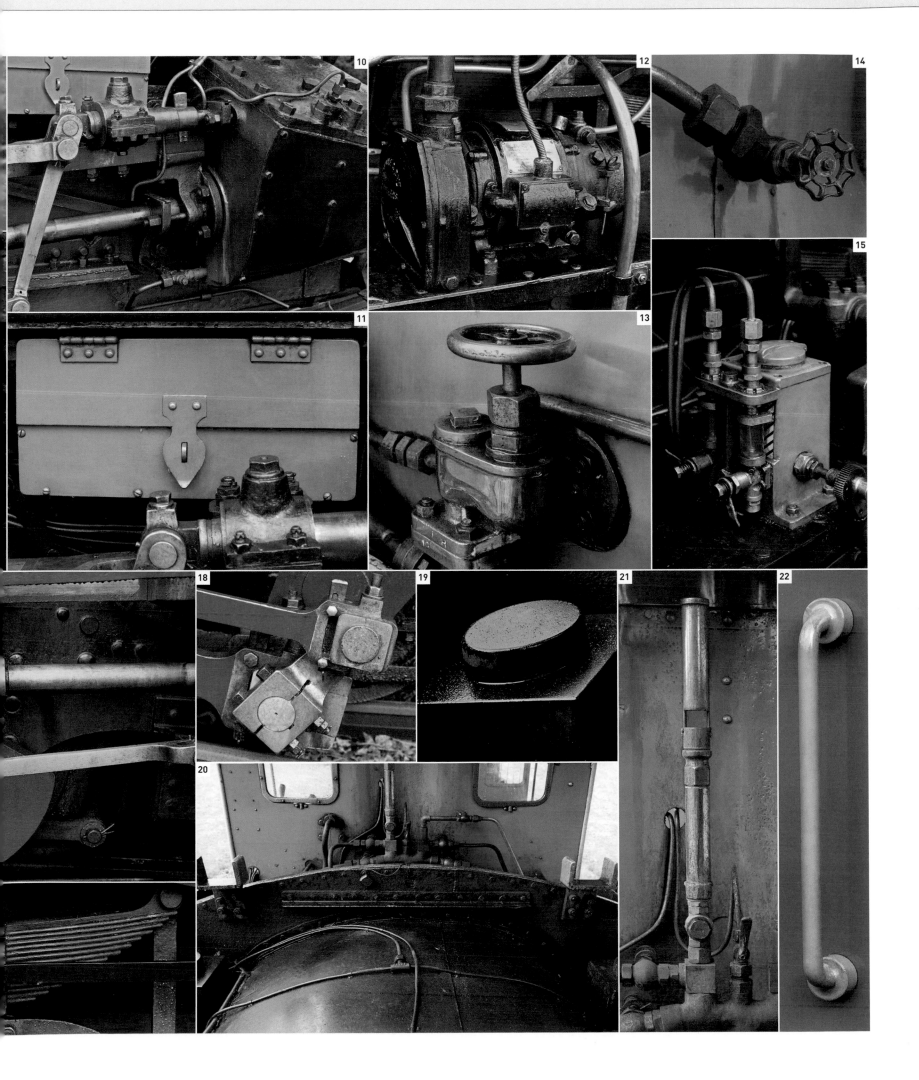

CAB INTERIOR

Driven from the right and fired from the left, the B Class travels uphill chimney-first on the DHR in India and is not turned around. As a result, the crew in the open cab tend to endure an unpleasant experience when the train runs downhill in poor weather. Since the DHR shares much of its route with the parallel cart road, the driver has to make frequent use of the whistle at the numerous crossings along the way.

1. Cab with firehole door at bottom, handbrake on left 2. Water level gauge for engine and tender tanks 3. Air reservoir gauge mounted on tender 4. Steam valves for "blower" (above) and driver's side injector (below) 5. Back of boiler with steam regulator 6. Boiler water level gauges 7. Boiler pressure gauge (left), steam chest gauge (right) 8. Vacuum brake valve 9. Reversing lever 10. Air brake valve 11. Doors to tender behind cab 12. Empty tender behind cab

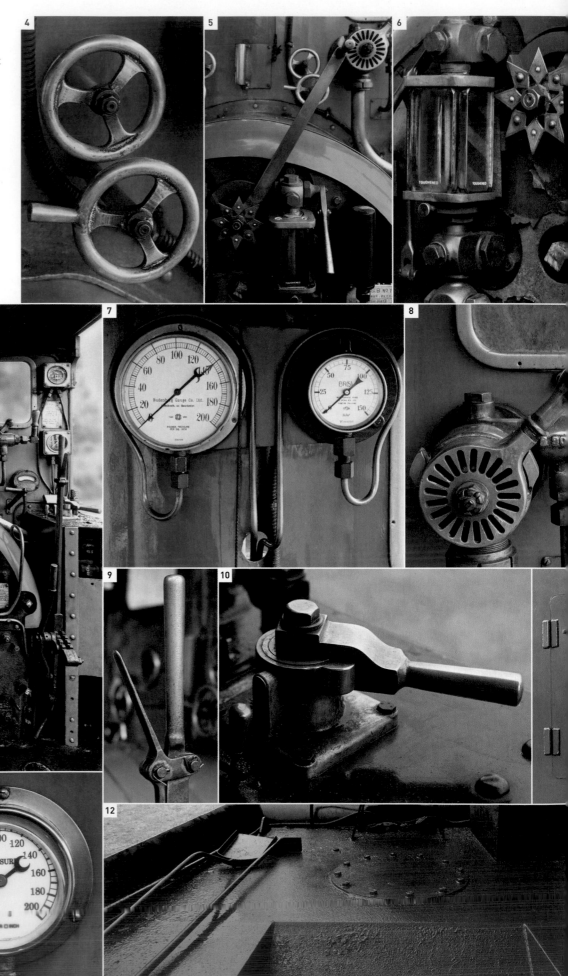

CARRIAGES

The carriages attached to No. 19 are replicas of carriages ordered for the DHR in 1967. One is a 29-seat saloon, the other a brake/saloon which contains a guard's compartment. The accommodation was reclassified second class when third class was abolished.

13. Internal view of first carriage 14. Ceiling light 15. Passenger emergency alarm 16. Loudspeaker 17. Warning, in Hindi, of fine for travelling without a ticket 18. Door handle 19. Metal pull to open window 20. Wooden seating 21. Internal view of brake carriage 22. Guard's van at rear of brake carriage 23. Guard's emergency vacuum brake 24. Light switches in guard's van 25. Guard's handbrake 26. Vacuum brake gauge 27. Air brake reservoir pressure gauge 28. Air brake pipe pressure gauge 29. Hindi and English script on outside of coach with acronym NF: Northeast Frontier Railway (India) 30. External view of door handle and handrail

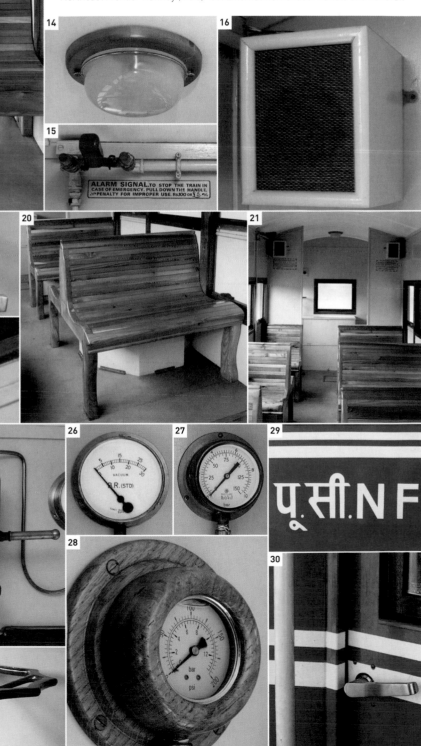

The First Electric Passenger Train

Although it may look like a ride at an amusement park, this train was the forerunner of every electric train that we see today. Developed by Werner von Siemens – a successful electrical engineer and a pioneer in the development of electric motors – it was unveiled in 1879 at a trade fair held in Berlin.

Earlier attempts at electric traction generated power within the locomotive, which limited the possibilites of rail travel. Siemens, however, established an alternative power source by drawing a continuous current (150 volts) for his 2.2-kilowatt motor from a conductor rail placed along the centre of the track. The train operated for four months and carried 90,000 people, despite advertising a top speed of just 4.4 mph (7 km/h) – though it is said to have reached 8.12 mph (13 km/h).

BUILDING ON SUCCESS

This experimental train set the future design for railways. Although energized lines were eventually superseded by safer and more efficient electric overhead wires, the success of the train enabled Siemens to develop an electric tramway, which began operating in the Lichterfelde district of Berlin in 1881. Both of these pioneering designs, created by Siemens with his partner, mechanical engineer Johann Halske, formed the bedrock of their worldwide electrical engineering business, which is still operating today.

Visitors to the 1879 Berlin trade fair were carried around a 984-ft (300-m) circular track by Siemens & Halske's electric train.

1895-1913
GOLDEN AGE

GOLDEN AGE

One of the world's oldest railways pointed the way forward when its first mainline electric route was opened in 1895. The Baltimore & Ohio Railroad, which dates back to 1830, installed electrification in its Howard Street Tunnel as a response to problems with locomotive fumes. Within 10 years an experimental electric railcar running on a military line snatched a new world speed record in Germany in 1903.

The period also saw the appearance of the compression ignition, oil-fuelled locomotive – a precursor of the mass move to diesel traction that followed later. But steam locomotives still had plenty of life, and engineers around the globe worked towards increasing their efficiency. In Britain, the Great Western Railway's George Jackson Churchward shaped the future of the country's steam traction when he came up with a new range of locomotives using standardized parts, having adapted ideas from overseas. As cities around the world grew, the craze for underground railways spread; the iconic Metro system in Paris and the Subway in New York were among those to begin passenger services during this era.

Engineering feats included the Victoria Falls Bridge across the Zambezi River in Africa, which opened in 1905; and the Simplon Tunnel, which opened in 1906. The structure, stretching more than 12 miles (20 km) under the Alps to connect Italy and Switzerland, became the world's longest tunnel.

△ **Stylish French Metro**
The entrances to the new Paris Metro, which opened in 1900, were inspired by the Art Nouveau movement of the period.

Key Events

▷ **1895** America's Baltimore & Ohio Railroad launches the electric age with an electrified route through the Howard Street Tunnel.

▷ **1896** Britain's first compression ignition oil locomotive is developed – the precursor of today's diesels.

▷ **1896** Budapest's first metro line is completed.

▷ **1900** The first section of the Paris Metro is opened.

▷ **1902** George Jackson Churchward's innovative 4-6-0 for the Great Western Railway helps change the direction of British locomotive design.

▷ **1902** Berlin's first *Untergrundbahn* underground line is finished.

▷ **1902** The New York Central Railroad launches the *20th Century Limited* express passenger train.

▷ **1903** A German experimental electric railcar reaches 131 mph (211 km/h).

▷ **1904** New York's Subway opens its first section.

▷ **1906** The Simplon Tunnel connects Italy and Switzerland.

△ **Jura-Simplon Railway**
A 1900 timetable for the Jura-Simplon Railway. In 1895 the company proposed the ambitious project for the building of the Simplon Tunnel.

▷ **1909** The first Beyer-Garratt articulated steam locomotive is completed.

▷ **1912** A mainline diesel goes on test for Germany's Prussian state railways.

> "**Railway termini** ... are our gates to **the glorious and the unknown.** Through them we pass out into **adventure and sunshine,** to them, alas! we return"

E.M. FORSTER, BRITISH AUTHOR

◁ **The Forth Bridge** created a direct route between London and Aberdeen, prompting a second "Race to the North" in 1895

Express Steam for the UK

This period of British railway history saw major advances in the design and construction of British express passenger steam locomotives. Innovations – often developed in other countries – such as compounding using high- and low-pressure cylinders, larger and higher pressure boilers, superheating, and longer wheel arrangements all contributed to more efficient locomotives. These graceful machines were able to haul longer and heavier trains at greater speeds on Britain's busy main lines.

△ MR Class 115, 1896

Wheel arrangement	4-2-2
Cylinders	2 (inside)
Boiler pressure	170 psi (11.95 kg/sq cm)
Driving wheel diameter	93 in (2,370 mm)
Top speed	approx. 90 mph (145 km/h)

These express locomotives, designed by Samuel W. Johnson, were built at the Midland Railway's Derby Works till 1899. Class 115s were nicknamed "Spinners" for the spinning motion of their pair of huge driving wheels.

△ GNR Class C2 Small Atlantic, 1898

Wheel arrangement	4-4-2
Cylinders	2
Boiler pressure	170 psi (11.95 kg/sq cm)
Driving wheel diameter	93 in (2,370 mm)
Top speed	approx. 90 mph (145 km/h)

Named *Henry Oakley*, No. 990 was the first of 22 C1 Class express locomotives designed by Henry Ivatt and built at the Great Northern Railway's Doncaster Works. Nicknamed "Klondyke", it was passed to the London & North Eastern Railway, which went on to classify this small boiler version as C2.

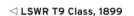

◁ LSWR T9 Class, 1899

Wheel arrangement	4-4-0
Cylinders	2 (inside)
Boiler pressure	175 psi (12.30 kg/sq cm)
Driving wheel diameter	79 in (2,000 mm)
Top speed	approx. 85 mph (137 km/h)

Nicknamed "Greyhounds", 66 T9 Class passenger locomotives were built between 1899 and 1901. The class was designed by Dugald Drummond for the London & South Western Railway.

◁ **MR Compound 1000 Class, 1902**

Wheel arrangement	4-4-0
Cylinders	3 (2 outside low-pressure; 1 inside high-pressure)
Boiler pressure	220 psi (15.46 kg/sq cm)
Driving wheel diameter	84 in (2,134 mm)
Top speed	approx. 85 mph (137 km/h)

Designed by Samuel W. Johnson, these express compound locomotives were built at the Midland Railway's Derby Works from 1902. Some 45 were constructed.

▷ **LNER Class C1 Large Atlantic, 1902**

Wheel arrangement	4-4-2
Cylinders	2
Boiler pressure	170 psi (11.95 kg/sq cm)
Driving wheel diameter	80 in (2,030 mm)
Top speed	approx. 90 mph (145 km/h)

Developed from the Great Northern Railway's Class C2 Small Atlantic, 94 of these large boiler express locomotives were built at Doncaster Works between 1902 and 1910. Under London & North Eastern Railway's ownership, it retained its C1 classification to distinguish it from its small boiler relatives.

▽ **GWR, 3700 Class or City Class, 1902**

Wheel arrangement	4-4-0
Cylinders	2 (inside)
Boiler pressure	200 psi (14.06 kg/sq cm)
Driving wheel diameter	80 in (2,030 mm)
Top speed	approx. 100 mph (161 km/h)

Designed by George Churchward, 20 of these express locomotives were built at the Great Western Railway's Swindon Works between 1902 and 1909. In 1904, No. 3440 *City of Truro* was claimed to be the first steam locomotive to reach 100 mph (161 km/h).

◁ **GWR 4000 Class or Star Class, 1907**

Wheel arrangement	4-6-0
Cylinders	4 (2 outside, 2 inside)
Boiler pressure	225 psi (15.82 kg/sq cm)
Driving wheel diameter	80 in (2,030 mm)
Top speed	approx. 90 mph (145 km/h)

Another of George Churchward's designs, 73 Star Class express passenger locomotives were built at the Great Western Railway's Swindon Works between 1907 and 1923. The prototype, No. 4, was given the name *North Star*, then renumbered 4000. This is No. 4005 *Polar Star*, which remained in service until 1934.

British Evolution

By the end of the 19th century Britain's railway network had expanded to serve nearly every part of the country. Coal mines, quarries, ironworks, factories, ports, and harbours were all connected to the railway system, and the rapid growth of freight traffic led to the development of more powerful steam locomotives capable of handling heavier and longer trains. These freight workhorses were so successful that many remained in service for more than 50 years. At the same time, passenger traffic connecting cities with their suburbs also saw a rapid expansion, with new types of tank locomotives capable of fast acceleration hauling commuter trains to tight schedules.

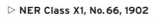

△ CR 812 Class, 1899

Wheel arrangement	0-6-0
Cylinders	2 (inside)
Boiler pressure	160 psi (11.25 kg/sq cm)
Driving wheel diameter	59³⁄₄in (1,520 mm)
Top speed	approx. 55 mph (88 km/h)

John F. McIntosh designed this tender locomotive for the Caledonian Railway. A total of 79 of the 812 Class were built between 1899 and 1909. Most remained in service for more than 50 years.

▷ NER Class X1, No. 66, 1902

Wheel arrangement	2-2-4T
Cylinders	2 (compound, inside)
Boiler pressure	175 psi (12.30 kg/sq cm)
Driving wheel diameter	67³⁄₄in (1,720 mm)
Top speed	approx. 55 mph (88 km/h)

Built for the North Eastern Railway in 1869 to haul its Mechanical Engineer's saloon, No. 66 *Aerolite* was rebuilt as a 4-2-2T in 1886 and as a 2-2-4T in 1902.

△ Met E Class No. 1, 1898

Wheel arrangement	0-4-4T
Cylinders	2 (inside)
Boiler pressure	150 psi (10.53 kg/sq cm)
Driving wheel diameter	65³⁄₄in (1,670 mm)
Top speed	approx. 60 mph (96 km/h)

No. 1 was the last locomotive built at the Metropolitan Railway's Neasden Works and spent its early years hauling commuter trains between Baker Street and Aylesbury. As London Transport No. L44, it remained in service until 1965 and is now preserved.

Shifting Freight

The railway companies built thousands of four-wheel covered and open freight wagons to carry raw materials, finished goods, and food perishables around Britain. Individual companies also owned large fleets of private-owner wagons and displayed their names on the sides. At docks and harbours, small tank locomotives with short wheelbases carried out shunting operations on the tightly curved railways.

◁ Alexandra Docks (Newport and South Wales) & Railway Co. No. 1340, 1897

Wheel arrangement	0-4-0ST
Cylinders	2
Boiler pressure	160 psi (11.25 kg/sq cm)
Driving wheel diameter	35³⁄₄in (910 mm)
Top speed	approx. 30 mph (48 km/h)

Built by the Avonside Engine Company of Bristol, this engine spent much of its life shunting around Newport Docks before being sold to a Staffordshire colliery in 1932. Now named *Trojan*, it is preserved at Didcot Railway Centre.

▷ GWR 2800 Class, 1903/1905

Wheel arrangement 2-8-0

Cylinders 2

Boiler pressure 225 psi (15.82 kg/sq cm)

Driving wheel diameter 55½ in (1,410 mm)

Top speed approx. 50 mph (80 km/h)

Eighty-four of these heavy freight locomotives, designed by George Churchward, were built at the Great Western Railway's Swindon Works between 1903 and 1919. Most were in service until the early 1960s.

◁ GWR Steam Railmotor, 1903

Wheel arrangement 0-4-0 + 4-wheel unpowered bogie

Cylinders 2

Boiler pressure 160 psi (11.25 kg/sq cm)

Driving wheel diameter 48 in (1,220 mm)

Top speed approx. 30 mph (48 km/h)

Built by the Great Western Railway, these self-propelled carriages were fitted with a steam-powered bogie and a vertical boiler at one end, and a driver's compartment at both ends. The railmotors operated suburban passenger services in London, and on country branch lines in England and Wales. A re-creation was completed by the Great Western Society in 2011 using an original body and a new power bogie.

△ LTSR Class 79, 1909

Wheel arrangement 4-4-2T

Cylinders 2

Boiler pressure 170 psi (11.95 kg/sq cm)

Driving wheel diameter 78 in (1,980 mm)

Top speed approx. 65 mph (105 km/h)

Four of these suburban tank engines, designed by Thomas Whitelegg, were built for the London, Tilbury & Southend Railway's commuter services from Fenchurch Street station in 1909. Retired in 1956, *Thundersley* is now part of the UK's national collection.

◁ GWR Iron Mink Covered Wagon, 1900

Type 4-wheel

Weight 10 tons (10.16 tonnes)

Construction iron

Railway Great Western Railway

More than 4,000 of these covered wagons were built by the Great Western Railway from 1886 to 1902. Ventilated and refrigerated versions carried meat, fish, and fruit. Bogie versions weighing 30 tons (30.5 tonnes) were built between 1902 and 1911.

△ The Royal Daylight Tank Wagon, 1912

Type 4-wheel

Weight 14 tons (14.2 tonnes)

Construction iron

Railway private owner

Built for the Anglo-American Oil Co. by Hurst Nelson of Motherwell, UK, this private-owner tank wagon carried imported American lamp oil branded as Royal Daylight. It is now displayed at Didcot Railway Centre.

GWR Auto Trailer No. 92

Great Western Railway's Auto Trailer No. 92, built at Swindon Works in the UK in 1912 and now based at Didcot Railway Centre, is a unique survivor of one of the earliest types of GWR "auto coach". It is essentially a passenger carriage with a built-in driving compartment at one end with controls that link to the steam railmotor to which it is coupled as a two-car unit. The ensemble can therefore be driven in either direction without the need for the locomotive to "run round" when it has reached its destination.

RESTORED TO ITS ORIGINAL GWR Crimson Lake livery, the 70-seater Auto Trailer No. 92 is the non-powered, trailing "half" of the Great Western Society's Railmotor & Trailer "set". The "powered half" is the railmotor itself (No. 93, pictured above), a near-identical timber-bodied vehicle that has its own built-in, vertical-boilered steam engine, and seating for 50 passengers. The two vehicles ran coupled together as a "steam multiple unit" – the ancestor of today's modern multiple unit trains – on GWR's branch lines and on their main lines as a "stopping" passenger train.

When operating railmotor first, the driver and fireman work in the engine compartment. When travelling auto trailer first, the fireman remains with the engine operating the valve gear and injectors, and feeding the fire, while the driver moves to a compartment at the front of the auto trailer. From there he has command of the unit's basic controls, which are connected to the engine by a series of interacting rods, linkages, pipes, or chains. He can also sound a warning bell on the front of the coach.

FRONT VIEW

REAR VIEW

SPECIFICATIONS FOR RAILMOTOR			
Class	Railmotor	In-service period	1912–57 (No. 93)
Wheel arrangement/cylinder	0-4-0 + 4-wheel bogie	Cylinders	2
Origin	UK	Boiler pressure	160 psi (11.25 kg/sq cm)
Designer/builder	George J. Churchward	Driving wheel diameter	48 in (1,220 mm)
Number produced	18 railmotors	Top speed	approx. 30 mph (48 km/h)

SPECIFICATIONS FOR AUTO TRAILER NO. 92	
Origin	UK
In-service	1912–57
Coaches	1 (couples with a railmotor)
Passenger capacity	70 seats (plus 50 in railmotor)
Route	Great Western Railway routes

Corridor connection to railmotor, which contains the engine

Luggage compartment at rear of auto trailer

Smoking saloon has capacity for 30 passengers

Central entrance vestibule with retractable steps

Non-smoking saloon seats 40 passengers

Driving compartment is used when the auto trailer is in front

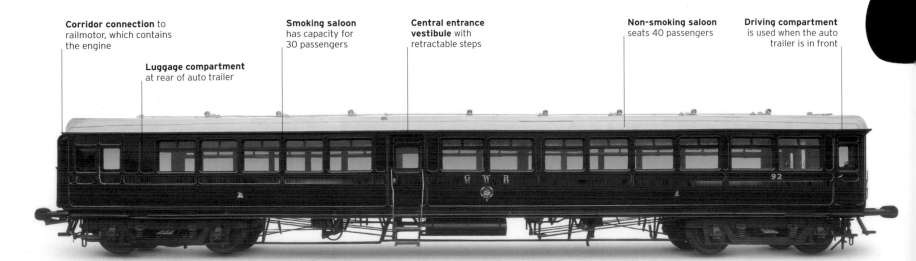

Driving compartment
From his forward-facing driving compartment at one end of the auto trailer the driver has command of a regulator lever and vacuum brake, which are connected to the steam railmotor, and a bell to signal to the guard and the fireman.

Great Western cities
The garter design for GWR's coat of arms, which includes the heraldic shields of the cities of London and Bristol, was adopted from 1870 and displayed extensively throughout their system.

EXTERIOR

In the early and later years of the GWR its coaches were all finished in a brown and cream livery, but in 1912–22 the railway standardized on a dark red, called Crimson Lake. Completed in 1912, the auto trailer was finished in this Crimson Lake livery with gold lining – some 1,200 ft (366 m) of it – and GWR insignia. The recent restoration project, completed in 2012, has returned the auto trailer to these original colours.

1. Carriage number 2. Driver's warning bell 3. Coat of arms of the City of Bristol 4. Destination board attached to side of carriage 5. Sign on luggage compartment door 6. Fold-down passenger steps into carriage 7. Brass door handle to passenger compartment 8. Pressure gauge on gas tank 9. Secondary suspension of transverse leaf springs 10. Part of bogie 11. Gas tank for carriage lighting 12. Rear buffer

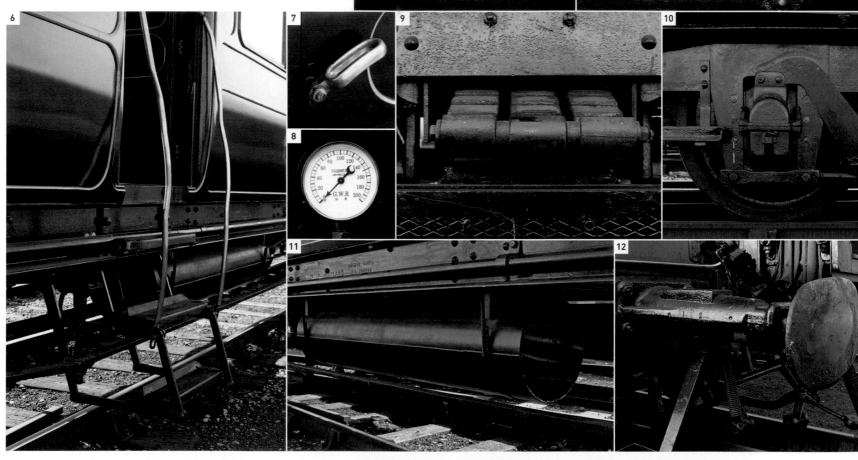

DRIVING COMPARTMENT

The spacious interior of the driving compartment gives the driver control of the train's basic controls. It also includes a fold-down seat – but this is rarely used as the driver has to stand to be able to reach and operate the regulator. Communication between the driver, fireman, and guard is via an electric (battery-powered) bell, and a series of simple bell codes: one ring for "start", two for "stop", and three for "brakes off". For the driver's comfort there is a steam-heat radiator, and there are windscreen wipers too – but 1912 technology did not extend to an electric motor to run them, so manual operation was necessary.

13. Cab interior with regulator lever above control window to allow driver to control the steam railmotor from the auto trailer 14. Vacuum gauge 15. Lever to open sandbox 16. Bell to signal to other members of train crew 17. Vacuum brake control 18. Foot treadle to sound exterior warning bell

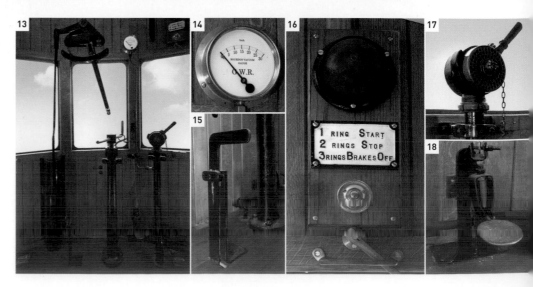

CARRIAGE INTERIOR

Restored by craftsmen at the Llangollen Railway in North Wales, the seating in No. 92's two passenger saloons is authentically upholstered in GWR-style, diamond-pattern brown moquette. Some of the seats featuring "flip over" backs, which allow passengers to face the direction of travel, were recovered from a derelict tramcar in Adelaide, Australia.

19. Overview of carriage interior **20.** Replica of original gas light fitting, now powered by electricity **21.** Roller blind **22.** Wooden, hand-carved corbel **23.** Electric light switches (a modern addition) **24.** Hand strap suspended from ceiling with decorative metal brackets **25.** Armrest between seats **26.** Part of heater under seats, fed with steam from the railmotor boiler **27.** Metal seat leg **28.** Smoking saloon sign on glass window **29.** Match striker in smoking compartment **30.** Leather strap to open and close window **31.** Emergency pull chain **32.** Decorative brass handles on door leading to carriage **33.** Ticket rack in guard's vestibule between passenger saloons **34.** Lever for releasing exterior fold-down steps **35.** Twin luggage doors **36.** Luggage door locking mechanism **37.** Wicket gate at end of carriage leads to railmotor compartment

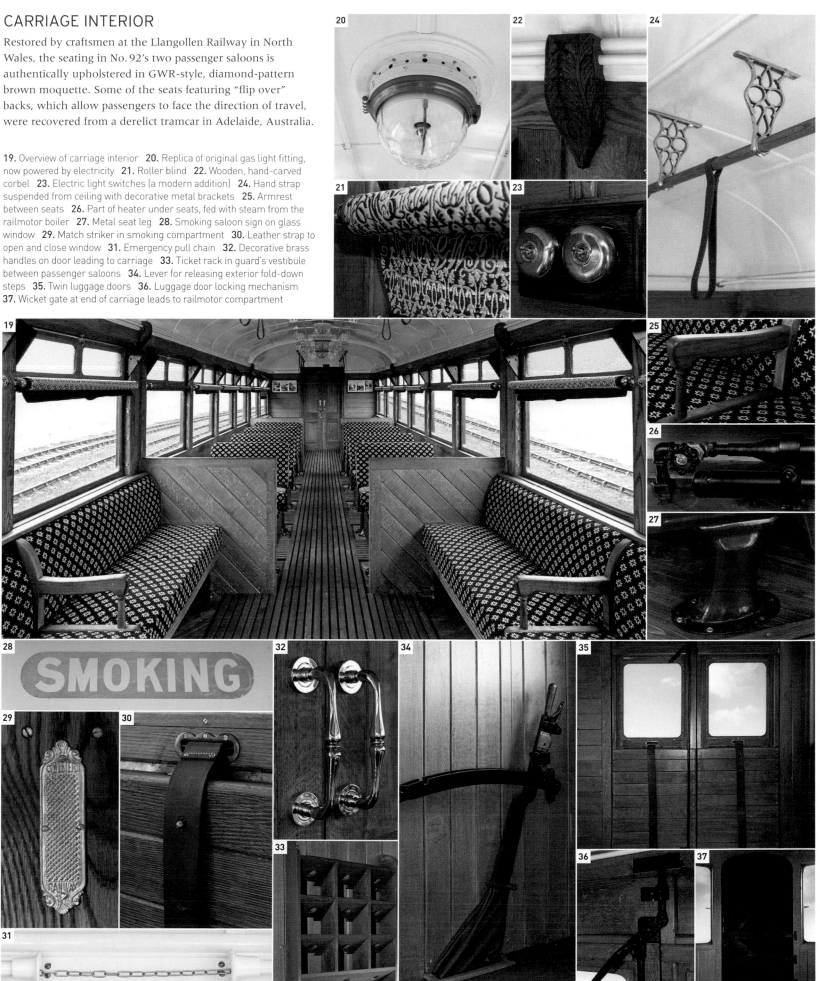

Continental Glamour

Railways had conquered most parts of Europe, and trains were now carrying vast quantities of raw materials and finished goods as well as large numbers of passengers. Travelling times between European cities had been cut significantly thanks to improvements in track and signalling, and also to modern coaches and powerful locomotives capable of sustaining higher speeds for greater lengths of time. New technology led the way as superheated and compound engines rolled off the production lines in ever greater numbers, while the US-influenced 4-6-2 "Pacific" type also started to make an appearance.

◁ **Nord Compound, 1907**

Wheel arrangement	4-6-0
Cylinders	4 (compound)
Boiler pressure	232 psi (16.3 kg/sq cm)
Driving wheel diameter	69 in (1,750 mm)
Top speed	approx. 70 mph (113 km/h)

French engineer Alfred de Glehn designed these compound express locomotives. Built for railways in France and abroad, some remained in service until the 1960s.

△ **Bavarian Class S3/6, 1908**

Wheel arrangement	4-6-0
Cylinders	4 (compound)
Boiler pressure	213 psi (15 kg/sq cm)
Driving wheel diameter	73½ in (1,870 mm)
Top speed	approx. 75 mph (120 km/h)

Designed by the German company Maffei, a total of 159 of these express locomotives were built over a period of nearly 25 years – 89 for the Royal Bavarian State Railways and 70 (known as Class 18.4-5) for the Deutsche Reichsbahn – between 1908 and 1931. This example was modernized in the 1950s.

1895 Paris Crash

On the afternoon of 22 October 1895 an express train from Granville hauling three baggage cars, a post van, and six passenger carriages approached the Montparnasse terminus, Paris. The train was travelling too fast, the air brake failed, and it crashed through the buffer stop at 30 mph (48 km/h), then travelled across the station concourse, through the station wall, and down to the street. A woman pedestrian was killed, but amazingly there were no fatalities on the train.

The infamous accident Locomotive No. 721 lies upended on its nose after crashing through the 2-ft- (60-cm-) thick wall of the terminus and falling 33 ft (10 m) onto the street below.

▽ **Prussian Class P8, 1908**

Wheel arrangement	4-6-0
Cylinders	2
Boiler pressure	170 psi (11.95 kg/sq cm)
Driving wheel diameter	69 in (1,750 mm)
Top speed	approx. 68 mph (110 km/h)

One of the most successful European steam locomotive designs, around 3,700 of the Prussian state railways superheated Class P8s were built between 1908 and 1926. Designed by Robert Garbe, they were built in several different German factories.

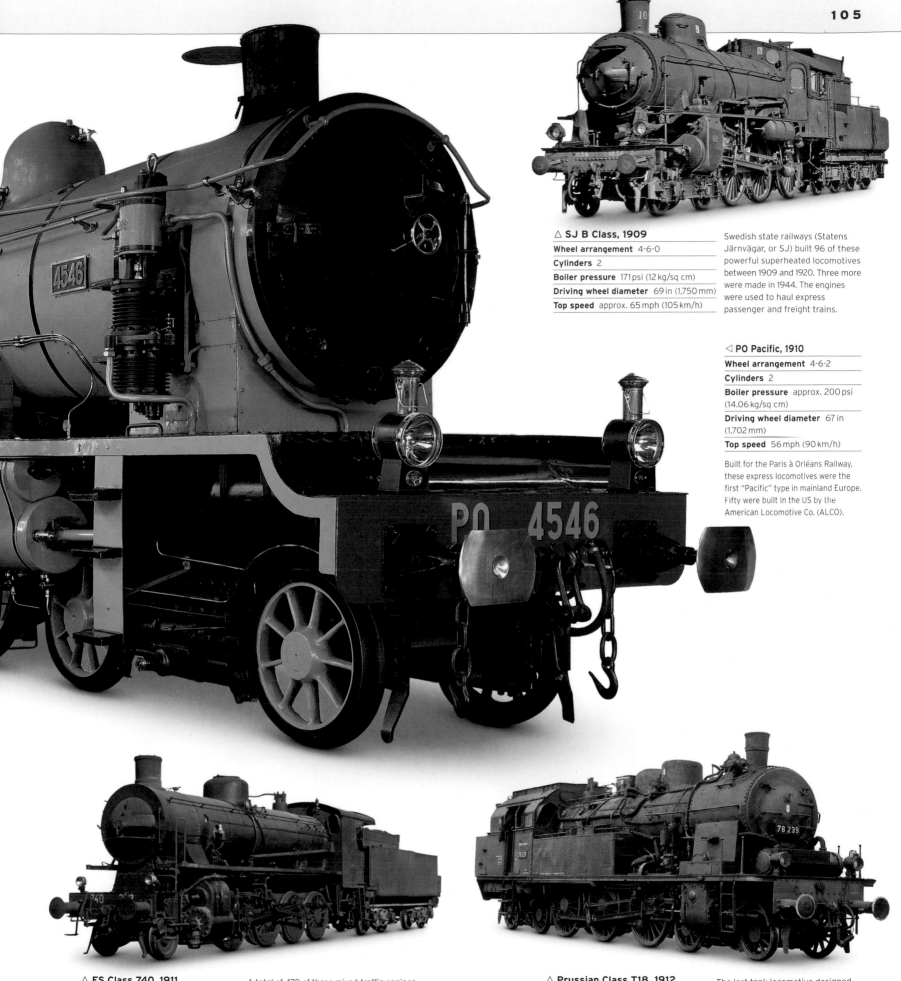

△ SJ B Class, 1909

Wheel arrangement	4-6-0
Cylinders	2
Boiler pressure	171 psi (12 kg/sq cm)
Driving wheel diameter	69 in (1,750 mm)
Top speed	approx. 65 mph (105 km/h)

Swedish state railways (Statens Järnvägar, or SJ) built 96 of these powerful superheated locomotives between 1909 and 1920. Three more were made in 1944. The engines were used to haul express passenger and freight trains.

◁ PO Pacific, 1910

Wheel arrangement	4-6-2
Cylinders	2
Boiler pressure	approx. 200 psi (14.06 kg/sq cm)
Driving wheel diameter	67 in (1,702 mm)
Top speed	56 mph (90 km/h)

Built for the Paris à Orléans Railway, these express locomotives were the first "Pacific" type in mainland Europe. Fifty were built in the US by the American Locomotive Co. (ALCO).

△ FS Class 740, 1911

Wheel arrangement	2-8-0
Cylinders	2
Boiler pressure	171 psi (12 kg/sq cm)
Driving wheel diameter	55 in (1,400 mm)
Top speed	approx. 56 mph (90 km/h)

A total of 470 of these mixed-traffic engines were built for the Italian state railways (Ferrovie dello Stato, or FS) between 1911 and 1923, some remaining in service until the 1970s. No. 740.423 has been restored to operational condition in Sardinia, and is occasionally used on charter trains.

△ Prussian Class T18, 1912

Wheel arrangement	4-6-4T
Cylinders	2
Boiler pressure	170 psi (11.95 kg/sq cm)
Driving wheel diameter	65 in (1,650 mm)
Top speed	approx. 62 mph (100 km/h)

The last tank locomotive designed for the Prussian state railways, 534 Class T18s were built between 1912 and 1927. Some were still in service in the 1970s with Deutsche Bundesbahn in West Germany and Deutsche Reichsbahn in East Germany.

Fulgence Bienvenüe
1852–1936

French civil engineer Fulgence Bienvenüe was the creator of the Paris Métro, a network that revolutionized the daily lives of Parisians. His extraordinary achievement followed an inauspicious beginning to his railroad career; in 1881 he lost his left arm in a construction accident while working on his first rail project in Normandy, France. However, this did not deter him from pursuing his engineering ambitions, and after moving to Paris in 1886, he became chief engineer for the Métro and supervised its development over the next 35 years. In addition to the Métro, Bienvenüe also managed engineering projects for the Parisian highway, lighting, and cleaning departments.

FATHER OF THE METRO

With Paris hosting the Universal Exhibition in 1900, the city's Municipal Council asked Bienvenüe to draw up plans for a narrow-gauge metro network for electric trains. The project started on 4 October 1898 and the first Métro line (Line 1, Porte de Vincennes to Porte Maillot) opened to passengers on 19 July 1900, in time for the exhibition.

The speed and efficiency of this new urban transport system impressed Parisians so much that the council granted Bienvenüe the job of extending and building a full underground network. Progress was swift. Within five years Lines 2 and 3, which stretched for 26 miles (42 km), were completed despite a number of unforeseen setbacks, including a fire at Couronnes in 1903 in which 84 people died. When Line 4 was tunnelled under the River Seine (1904–10), the construction techniques used were hailed as master strokes of civil engineering. By the eve of World War I, the Paris Métro was largely complete.

In 1933 the Avenue du Maine station was renamed Bienvenüe in honour of the "father of the Paris Métro". Nowadays, with some 1.5 billion journeys made on the Métro each year, the network is an integral part of the city.

Early Paris Métro
Three Métro lines (3, 7, and 8) cross one another beneath the Place de l'Opéra. The enormous construction effort to build the Métro saw the streets of central Paris torn up, much to the alarm of Parisians.

Honouring history
A Sprague-Thomson electric train arrives at Place de la Bastille Métro station on Line 1 in 1912. Paris Métro stations are named after significant events, places, and people from French history.

H&BT Caboose No. 16

Built by the Pennsylvania Railroad in 1913, this wooden, four-wheeled caboose, or cabin car, saw service on the railway's Middle Division between Harrisburg and Altoona before being sold to the Huntingdon & Broad Top Mountain Railroad & Coal Company (H&BT). Known as "bobbers", these cabooses were attached to the rear end of a freight train, serving as an office, lookout, and home for the crew during trips.

WIDELY USED ON NORTH AMERICAN railways from the 1870s through to the 1930s, "bobbers" got their nickname from railway crews for their bumpy and occasionally unstable riding conditions. The cupola on the roof offered all-round visibility for conductors, allowing them to watch the freight wagons during their journeys. Originally numbered No. 478396, this caboose was built at the Pennsylvania Railroad's Car Shops in Altoona and remained in service until 1940 when it was sold to the Huntingdon & Broad Top Mountain Railroad & Coal Company. It was then renumbered No. 16 on this coal-carrying short line in south central Pennsylvania and was one of the last wooden-bodied, four-wheeled "bobbers" to remain in service in the US before the railway's closure in 1954. Saved from the scrapyard, it then had several owners before it was donated to the Railroad Museum of Pennsylvania in 1998. Here it was expertly restored and is currently on display in H&BT red livery.

REAR VIEW

FRONT VIEW

H.&B.T.

Bright bobber
Restored caboose No. 16 now carries the initials of the Huntingdon & Broad Top Mountain Railroad (H&BT), which originally opened in 1855 to serve coal mines in Pennsylvania.

SPECIFICATIONS			
Type	Caboose (cabin car)	In-service period	1913–54
Origin	USA	Passenger capacity	1 conductor, crew's quarters
Designer/builder	Altoona Car Shops	Weight	12½ tons (12.7 tonnes)
Number produced	Not known	Railway	Pennsylvania Railroad/H&BT

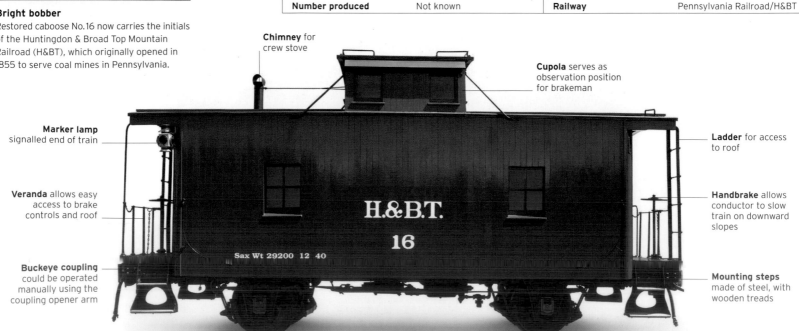

Chimney for crew stove

Cupola serves as observation position for brakeman

Marker lamp signalled end of train

Ladder for access to roof

Veranda allows easy access to brake controls and roof

Handbrake allows conductor to slow train on downward slopes

Buckeye coupling could be operated manually using the coupling opener arm

Mounting steps made of steel, with wooden treads

Crew's caboose
North American cabooses traditionally had a veranda at each end, which was reached by steps from ground level. A steel ladder enabled freight train crew to access the roof in order to clean the windows of the lookout cupola.

16

Sax Wt 29200 12 46

EXTERIOR

The caboose had a cupola from which the brakeman kept a lookout for overheating axles, or hotboxes, as well as shifting cargo and damage to the train. The buckeye coupling could be manually operated; use of this type of coupling and air brakes were made mandatory by US Congress in 1893, significantly reducing the number of railway accidents and workers killed or injured during coupling operations.

1. Caboose number painted on side 2. Coupling link
3. Coupling opener arm 4. Steps up to veranda
5. Marker lamp 6. Retaining valve to keep air brakes applied on long downward slopes 7. Whistle 8. Chimney
9. Windows in cupola 10. Wheel unit 11. Open journal box showing bearing 12. Brake wheel on platform

INTERIOR

The caboose's cosy interior was an office and a temporary home to the locomotive crew and conductor. Raised seats allowed views through the roof cupola, and a coal-fired stove bolted to a steel plate on the floor kept the crew warm at night and provided cooking facilities. Surrounded by protective steel plates, the stove was fitted with safety features such as a double-latched door to prevent hot coals spilling out, and a lip on the top to stop pans and pots from sliding off when the train was in motion.

13. Interior of caboose **14.** Window latch **15.** Air brake pressure gauge **16.** Seats in cupola **17.** Oil lamp **18.** Air controls on coal stove **19.** Coal stove **20.** Sink unit

Rapid Development

With railways now well established, this period saw rapid developments in the design of both passenger and freight locomotives around the world. Mass production of heavy freight engines reached new heights with more than 1,000 of the Prussian state railways Class G8 along with another 5,000 of the later Class G8.1 being built over the following years. However, the world record for the most numerous class of locomotive goes to the Russian E Class, of which around 11,000 were built.

▷ Austrian Gölsdorf Class 170, 1897

Wheel arrangement	2-8-0
Cylinders	2 (compound)
Boiler pressure	185 psi (13 kg/sq cm)
Driving wheel diameter	49$\frac{1}{2}$in (1,260 mm)
Top speed	approx. 37 mph (60 km/h)

Designed by Karl Gölsdorf for the Imperial Royal Austrian State Railways, the Class 170 freight locomotives were the first to be fitted with radially sliding coupled axles, known as Gölsdorf axles.

△ Prussian Class G8, 1902

Wheel arrangement	0-8-0
Cylinders	2
Boiler pressure	170 psi (11.95 kg/sq cm)
Driving wheel diameter	53 in (1,350 mm)
Top speed	approx. 35 mph (56 km/h)

More than 1,000 of these superheated freight locomotives were built in Germany for the Prussian state railways. After WWI hundreds were given to Germany's enemies as reparations. Some saw service during the building of the Baghdad Railway in Turkey in 1916.

△ PRR Class E7, 1902

Wheel arrangement	4-4-2
Cylinders	2
Boiler pressure	205 psi (14.4 kg/sq cm)
Driving wheel diameter	78$\frac{1}{2}$in (2,000 mm)
Top speed	approx. 80 mph (129 km/h)

The original Class E7 No. 7002 was built at the Pennsylvania Railroad's Altoona Works, Pennsylvania, US. It was once claimed to be the world's fastest steam engine, supposedly reaching 127 mph (204 km/h), but this is disputed. First numbered 8063, this locomotive was renumbered after the first 7002 was scrapped and is now in the Pennsylvania Railroad Museum.

◁ Indian Class EM, 1907

Wheel arrangement	4-4-2
Cylinders	2
Boiler pressure	190 psi (13.4 kg/sq cm)
Driving wheel diameter	78 in (1,980 mm)
Top speed	approx. 60 mph (96 km/h)

Originally built as a 4-4-0 by the North British Locomotive Co. for the Great Indian Peninsula Railway, the Class EM remained in service until the late 1970s. EM No. 922 was rebuilt in 1941 by the Mughalpura workshops.

△ VGN Class SA, 1910

Wheel arrangement	0-8-0
Cylinders	2
Boiler pressure	200 psi (14.06 kg/sq cm)
Driving wheel diameter	51 in (1,295 mm)
Top speed	approx. 10 mph (16 km/h)

One of only five Class SA switcher locomotives built, Nos. 1, 2, and 3 were made at American Locomotive Co. (ALCO); Nos. 4 and 5 by Baldwin Locomotive Works. No. 4 (shown here) retired in 1957 as the last steam locomotive on the Virginian Railway.

▽ Russian E Class, 1912

Wheel arrangement	0-10-0
Cylinders	2
Boiler pressure	170 psi (11.95 kg/sq cm)
Driving wheel diameter	48 in (1,220 mm)
Top speed	approx. 30 mph (48 km/h)

First built at Lugansk Works in Ukraine, a large number of these heavy freight engines were eventually constructed in Russia, as well as in Czechoslovakia, Germany, Sweden, Hungary, and Poland. There were several subclasses, some of which were fitted with condensing tenders for working in areas where water was scarce.

△ Austrian Gölsdorf Class 310, 1911

Wheel arrangement	2-6-4
Cylinders	4 (compound)
Boiler pressure	220 psi (15.5 kg/sq cm)
Driving wheel diameter	84¼ in (2,140 mm)
Top speed	approx. 62 mph (100 km/h)

Designed by Karl Gölsdorf, 90 of the Class 310 four-cylinder compound express locomotives were built for the Imperial Royal Austrian State Railways from 1911 to 1916. This was one of the most elegant locomotives of the period.

TECHNOLOGY

Geared Locomotives

US-built, lighter-weight geared steam locomotives such as the Shay, Heisler, and Climax types had wheels driven by reduction gearing. These locomotives were designed for the quick and cheap-to-lay industrial railways used by logging, sugar-cane, mining, and quarrying industry operations where speed was not needed and gradients were often steep.

Heisler 2-truck geared locomotive No. 4 This locomotive, designed by Charles L. Heisler, was built for the Chicago Mill & Lumber Co. in 1918. It was the fastest of this type and is on display at the Railroad Museum of Pennsylvania.

VGN Class SA No. 4

One of only five Class SA 0-8-0 switchers (known as shunters in the UK), this powerful locomotive was delivered by the Baldwin Locomotive Works of Eddystone, Pennsylvania, to the newly formed Virginian Railway (VGN) in August 1910. It marshalled heavy coal trains at the railway's yards in Virginia and West Virginia until its retirement in 1957, when it was replaced by diesel locomotives. It is currently on display at the Virginia Museum of Transportation in Roanoke, and is the last surviving steam engine of the Virginian Railway.

OPENED IN 1909, THE VIRGINIAN RAILWAY became a highly profitable company by transporting high-quality coal from the mines in West Virginia to its piers at Sewells Point, Norfolk, southwestern Virginia, from where it was it was transferred on to ships. Nicknamed the "Richest Little Railroad in the World", the railway used some of the world's most powerful steam locomotives to haul its heavy eastbound coal trains up the steeply graded line to Clark's Gap in West Virginia, until this section of the railway was electrified in 1925.

Marshalling the long coal trains in Page (named after one of the railway's founders), and other yards in West Virginia and Virginia, was carried out by powerful 0-8-0 Class SA switchers, of which No. 4 is the only surviving example. Of the five Class SA switchers built, Nos. 1–3 were supplied by ALCo and Nos. 4–5 by the Baldwin Locomotive Works.

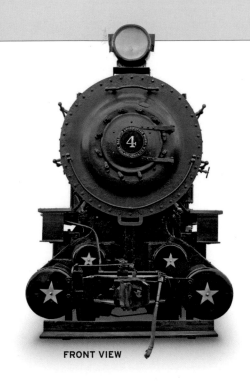

FRONT VIEW

REAR VIEW

Baldwin Locomotive Works
Founded by Matthias Baldwin in 1825, the Baldwin Locomotive Works built more than 70,000 engines for railways around the world. In 1956 production ceased after it lost out on a large order to supply diesels for the Pennsylvania Railroad.

SPECIFICATIONS			
Class	SA	In-service period	1910-57
Wheel arrangement	0-8-0	Cylinders	2
Origin	USA	Boiler pressure	200 psi (14.06 kg/sq cm)
Designer/builder	Baldwin Locomotive Works	Driving wheel diameter	51 in (1,295 mm)
Number produced	5	Top speed	approx. 10 mph (16 km/h)

Tender has a water capacity of 5,000 gallons (18,927 litres)

Coal bunker can hold 10 tons (10 tonnes)

Rear sand dome sands track behind driving wheels when reversing

Steam dome contains throttle

Front sand dome sands track ahead of driving wheels when going forwards

Powerful switcher
A utilitarian machine, Class SA No. 4 weighed in at 81 tons (82 tonnes), and had a tractive effort of 45,200 lb (20.502 kg). The rear eight-wheeled tender weighed almost 50 tons (50.8 tonnes) when fully loaded.

EXTERIOR

SA No. 4 was built as a utilitarian workhorse able to shont heavy coal trains at slow speeds in marshalling yards. It was fitted with "knuckle" couplings and a Westinghouse air brake, both US standard systems. The two air reservoirs for the brakes were housed between the two cylinders at the front of the locomotive.

1. Engine number on side **2.** Headlight **3.** Front coupler **4.** Valve chamber head with metal star detail **5.** Builder's plate on side of engine **6.** Brass bell on top of engine **7.** Whistle attached to steam dome **8.** Safety valves **9.** Piston rod **10.** Crosshead support yoke **11.** Driving wheels **12.** Driving wheel springs **13.** Steps leading to cab **14.** Exterior of cab with bright red window frames **15.** Tender behind cab **16.** Signage displaying tender water capacity **17.** Light on tender **18.** Handrail around edge of tender

CAB INTERIOR

The driver and fireman of SA No. 4 worked in a hot and uncomfortable cab. The driver was seated on the right-hand side, where he could see the road ahead and control the throttle and air brake. Unlike many American locomotives, which were fitted with a mechanical stoker, the humble switchers had to be manually fed coal from the tender into the firebox by the fireman using a large shovel.

19. Boiler backhead in cab 20. Engine and train brake 21. Steam pressure gauge 22. Interior of firebox 23. Air brake gauge
24. Auxiliary controls 25. Throttle lever (regulator) 26. Control valves
27. Control pedal 28. Driver's seat

The New York Elevated Railway

While London and Paris burrowed underground to meet the demand for a fast and reliable public transport system, New York chose the overground route. Between 1840 and 1870, the city's population had grown by more than half-a-million inhabitants. This increase overwhelmed the capacity of its horse-drawn bus and streetcar routes, several of which ran along the main avenues. Although it was considered unsafe and impractical to replace horses with steam engines, two local entrepreneurs, Charles Harvey and Rufus Gilbert, believed their locomotives could run on viaducts built over the streets. They introduced two elevated lines to the west of Manhattan Island before financial problems forced the authorities to take over the project.

Under the 1875 Rapid Transit Act, four lines were constructed, and these would form the heart of the New York Elevated Railway (or the "El" as New Yorkers called it). The routes ran northwards along Second, Third, Sixth, and Ninth Avenues, and further lines were added up to 1917.

Although the smoke and noise of steam locomotives had been supplanted by electric traction, by the late 1930s the "El" was considered outdated. The lines were demolished between 1938 and 1955 to make way for the New York Subway system.

Passengers ride behind a lightweight, Forney tank locomotive of the Third Avenue Elevated Railroad in 1896, above the wagons and streetcars of Bowery.

On Other Gauges

George Stephenson introduced the 4-ft 8½-in- (1.435-m-) gauge for British railways in 1830 and before long it became the standard gauge for many railways around the world. However, there were, and still are, many exceptions. In India a broader gauge of 5 ft 6 in (1.67 m) was used for many mainline railways, but more lightly laid lines had narrower gauges of 3 ft 3 in (1 m) or, for mountain railways, only 2 ft (0.61 m). While the standard gauge was usually the norm in mainland Europe and the US, there was also widespread use of narrow gauges in mountainous regions. The most extensive narrow-gauge network in the US was the Denver & Rio Grande Railroad's 3-ft- (0.91-m-) gauge system in Arizona, Utah, and New Mexico.

▷ NWE Mallet, 1897

Wheel arrangement	0-4-4-0
Cylinders	4
Boiler pressure	200 psi (14 kg/sq cm)
Driving wheel diameter	39½ in (1,000 mm)
Top speed	approx. 18 mph (30 km/h)

This engine was one of 12 powerful articulated steam locomotives built for the 3-ft 3-in- (1-m-) gauge Nordhausen-Wernigerode Railway in Germany. Several were lost in WWI but three are now with the NWE's successor the Harzer Schmalspurbahnen on the Harz Mountains in central Germany.

△ NWR ST, 1904

Wheel arrangement	0-6-2T
Cylinders	2 (inside)
Boiler pressure	150 psi (10.53 kg/sq cm)
Driving wheel diameter	51 in (1,295 mm)
Top speed	approx. 30 mph (48 km/h)

One of the first locomotives built at India's North Western Railway's Mughalpura Workshops, ST No. 707 was made from parts supplied by North British Locomotive Co. of Glasgow. Weighing 55 tons (55 tonnes), this 5-ft 6-in- (1.67-m-) gauge locomotive was employed for shunting duties. It is now on display at the National Rail Museum, New Delhi.

△ KS Wren Class, 1905

Wheel arrangement	0-4-0
Cylinders	2
Boiler pressure	140 psi (9.84 kg/sq cm)
Driving wheel diameter	20 in (500 mm)
Top speed	approx. 15 mph (24 km/h)

A total of 163 of these narrow-gauge locomotives were built by the British company Kerr Stuart for use on industrial railways around the world between 1905 and 1930. However, *Jennie* was made in 2008 for the 2-ft- (0.60-m-) gauge Amerton Railway, Staffordshire, by the Hunslet Engine Co.

▷ Indian SPS, 1903

Wheel arrangement	4-4-0
Cylinders	2 (inside)
Boiler pressure	160 psi (11.25 kg/sq cm)
Driving wheel diameter	78 in (1,980 mm)
Top speed	approx. 50 mph (80 km/h)

A range of standard designs was introduced for India, including the Standard Passenger (SP); when superheating was added it became the SPS. British designed, some of these engines had extremely long working lives. After partition in 1947, this one ran on the new Pakistan Railways until the 1980s.

△ Mh 399, 1906

Wheel arrangement	0-8+4
Cylinders	2
Boiler pressure	180 psi (12.65 kg/sq cm)
Driving wheel diameter	36 in (910 mm)
Top speed	approx. 25 mph (40 km/h)

Built by Krauss of Linz, this locomotive was made for the Austrian Railways' 2-ft 6-in- (0.76-m-) narrow-gauge Mariazell Railway. It had rear wheels that are also driven by coupling rods. Seen here is No. 399.06 preserved on the Mariazellerbahn, Austria.

△ TGR K Class Garratt, 1909

Wheel arrangement	0-4-0+0-4-0
Cylinders	4
Boiler pressure	195 psi (13.70 kg/sq cm)
Driving wheel diameter	31½ in (800 mm)
Top speed	approx. 25 mph (40 km/h)

The world's first Garratt-type articulated steam locomotive, No. K1 was built by Beyer Peacock & Co. of Manchester, England, for the Tasmanian Government Railway, Australia. It ran on the 2-ft- (0.60-m-) gauge North East Dundas Tramway. This historic locomotive was returned to Britain in 1947 and now hauls trains on the Welsh Highland Railway.

△ EIR No. 1354 *Phoenix*, 1907

Wheel arrangement	0-4-0WT
Cylinders	2 (inside)
Boiler pressure	120 psi (8.44 kg/sq cm)
Driving wheel diameter	36 in (910 mm)
Top speed	approx. 20 mph (32 km/h)

One of five railmotors built in England by Nasmyth Wilson & Company, *Phoenix* was made for the 5-ft 6-in- (1.67-m-) East Indian Railway in 1907. Later, in 1925, the coaches were removed and *Phoenix* was rebuilt in India as a small shunting engine. It is now on display at the National Rail Museum, New Delhi.

▷ Lima Class C Shay, 1906

Wheel arrangement	B-B-B
Cylinders	3
Boiler pressure	200 psi (14.06 kg/sq cm)
Driving wheel diameter	36 in (910 mm)
Top speed	approx. 15 mph (24 km/h)

Designed by US inventor Ephraim Shay, the Class C geared three-truck steam locomotive was first introduced in 1885. This Shay No. 1 was built by the Lima Locomotive & Machine Co. for a standard-gauge logging railroad in Pennsylvania in 1906. It can be seen at the Railroad Museum of Pennsylvania, Strasburg.

Building Great Railways
Trans-Siberian Railway

Crossing eight time zones, the 5,772-mile (9,289-km) Trans-Siberian Railway is the longest continuous railway line in the world. Extending from the Russian capital, Moscow, to Vladivostok on the Pacific coast, it provides a strategic route connecting Asia with Europe.

Kama River, near Perm
A metal-truss railway bridge straddles the Kama River in this early colour photograph from c. 1909–15. The Trans-Siberian crosses numerous major rivers.

BY 1890 THE RUSSIAN EMPIRE stretched east from its European borders, across the Ural Mountains and the vastness of Siberia, to the Pacific coast. While European Russia, west of the Urals, had experienced industrial growth and acquired railways in the 19th century (the first railway, opened in 1851, was between Moscow and St Petersburg), the lands to the east remained virtually untapped. With few roads into the region, the only means of transport were the mighty Siberian river systems, but these were only navigable for around five months of each year – for the remaining months they were frozen. A railway was the key to opening up this vast hinterland.

Construction of the government-funded Trans-Siberian Railway began in 1891 with the blessing of Tsar Alexander III and his son, the

Steaming around Lake Baikal
Golden Eagle Trans-Siberian Express is one of the luxury trains that runs the route. Less pampered journeys can be taken on a variety of domestic and international services.

future Tsar Nicholas II. Work began at both ends – Moscow and Vladivostok – with Russian soldiers and convicts employed as railway navvies. Progress was fast, and by 1898 the line stretched 3,222 miles (5,185 km) from Moscow to Irkutsk, near the western shore of Lake Baikal.

Further east, the line running from Vladivostok to Khabarovsk had already opened in 1897. However, the Amur line running west from Khabarovsk to Chita would not open until much later. Presented with difficult terrain in this region, a shortcut

RAILWAY EMBLEM

R U S S I A N   F E D E R A T I O N

Kirov

Yaroslavl'

Perm'

Moscow

Yekaterinburg

River crossings On its long route across Russia the railway crosses many great rivers, including the world's fifth-longest river, the Ob, at Omsk.

Krasnoyarsk

Omsk

Novosibirsk

BELARUS

3 Moscow
Russia's capital city, Moscow, is shown in 1890, a year before construction on the Trans-Siberian began. Trans-Siberian passenger trains for Vladivostok depart from Yaroslavsky Station, which was opened in 1904.

2 Construction near Yekaterinburg
The Trans-Siberian Railway was built at the rapid rate of 2 ½ miles (4 km) a day in summer conditions. To reduce costs, lighter rails were used than those standard in Europe.

UKRAINE

M O N

KAZAKHSTAN

KEY
● Start/Finish
● Main stations
||||| Main route
||||| Original route 1903
||||| Trans-Mongolian route
||||| Trans-Manchurian route

C H I N A

linking Vladivostok to Chita via Manchuria was built. However, following conflicts with Japan over Manchurian interests, a route on Russian soil was needed and work on the Amur line began.

Meanwhile, the eastern and western sections of the Trans-Siberian Railway had come to an end on opposite shores of Lake Baikal – at 5,387ft (1,642m) the deepest freshwater lake in the world. A train ferry, the ice-breaker SS *Baikal*, was launched in 1899 to carry complete trains across the lake. It could carry up to 24 railway carriages and a locomotive. The ferry service became redundant in 1905 when the Circum-Baikal Line opened around the rocky western shores of Lake Baikal – its 33 tunnels and 200 bridges were built by convicts and political prisoners at great cost to the state.

The Khabarovsk Bridge over the Amur River was built in 1913 and, with the Amur section completed in 1916, the entire line was opened.

Siberian landscape
Full electrification of the Trans-Siberian was completed in 2002. This earlier passenger train hauled by three diesel-electric locomotives heads through the empty landscape.

KEY FACTS

DATES
1891 Building begins from Vladivostok (east) and Moscow (west) towards the centre
1903 Original route via Manchuria is completed
1904 Circum-Baikal around Lake Baikal is finished
1916 Final route is completed and line opens

TRAINS
Train No. 002 *Rossiya* travels eastbound Moscow-Vladivostok; No. 001 runs westbound. A range of domestic Russian trains or direct international trains run from Moscow to Ulan Bator, Mongolia; Beijing, China; and Pyongyang, North Korea. Trains are Russian or Chinese rolling stock, depending on final destination. Luxury trains, such as the steam-hauled *Golden Eagle* and *Tsar's Gold* also run.

JOURNEY
Moscow to Vladivostok
5,772 miles (9,289 km); 6 days, 4 hrs, Train No. 002M

RAILWAY
Gauge Broad 4 ft 11 ⁵⁄₆ in (1.52 m)
Tunnels 33 on Circum-Baikal section; longest passenger tunnel Tarmanchukan, 1.4 miles (2.2 km)
Bridges Track crosses 16 major rivers, including the Volga, Ob, Yenisey, and Oka; the Khabarovsk Bridge over the Amur is longest at 8,500 ft (2,590 m)
Highest point 3,412 ft (1,040 m) at the Yablonovy Mountain pass near Chita
Lowest temperature -79½°F (-62°C) between Mogocha and Skovordino on the Amur section

HIGH-SPEED CONSTRUCTION

An amazing feat of human effort, the Trans-Siberian Railway was built by thousands of Russian soldiers, as well as convicts and political prisoners serving sentences of hard labour. After 25 years of construction, its completion fulfilled the dreams of Russia's last tsar.

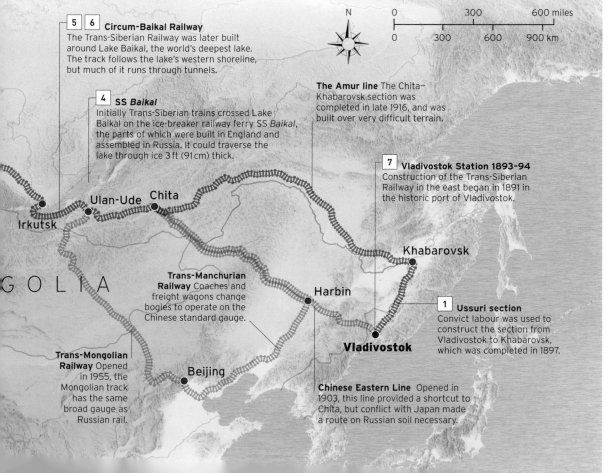

5 6 Circum-Baikal Railway
The Trans-Siberian Railway was later built around Lake Baikal, the world's deepest lake. The track follows the lake's western shoreline, but much of it runs through tunnels.

4 SS Baikal
Initially Trans-Siberian trains crossed Lake Baikal on the ice-breaker railway ferry SS Baikal, the parts of which were built in England and assembled in Russia. It could traverse the lake through ice 3 ft (91 cm) thick.

The Amur line The Chita–Khabarovsk section was completed in late 1916, and was built over very difficult terrain.

7 Vladivostok Station 1893-94
Construction of the Trans-Siberian Railway in the east began in 1891 in the historic port of Vladivostok.

Trans-Manchurian Railway Coaches and freight wagons change bogies to operate on the Chinese standard gauge.

1 Ussuri section
Convict labour was used to construct the section from Vladivostok to Khabarovsk, which was completed in 1897.

Trans-Mongolian Railway Opened in 1955, the Mongolian track has the same broad gauge as Russian rail.

Chinese Eastern Line Opened in 1903, this line provided a shortcut to Chita, but conflict with Japan made a route on Russian soil necessary.

Competition From the New Electrics

While steam traction was enjoying its heyday in the late 19th and early 20th centuries other forms of faster and cleaner rail transport were being developed. Electric trams, or streetcars, first started appearing in Europe and the US during the 1880s, and the technology began to appear on railways by the early 20th century. Using a mixture of either third-rail or overhead catenary power supplies, electric traction had been introduced on many city commuter lines in the UK and the US by the outbreak of World War I. With their fast acceleration these trains were ideal for lines with high-density traffic; they also eliminated the problem of pollution in built-up areas and in tunnels. In the US the electrification of the 2³/₄-mile (4.23-km) Cascade Tunnel in Washington State in 1909 was an early example of clean electric locomotives replacing the asphyxiating fumes of steam engines in confined spaces.

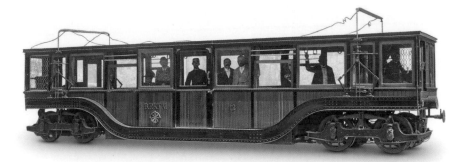

△ Budapest Metro car, 1896

Wheel arrangement	2 x 4-wheel powered bogies with 28 PS motors
Power supply	300 V DC, overhead supply
Power rating	28 hp (20.59 kW) per engine
Top speed	approx. 30 mph (48 km/h)

Fitted with two Siemens & Halske traction motors, 20 of these double-ended, electric subway cars were built for Continental Europe's first electric underground railway, which opened in Budapest, Hungary in 1896. Plans for extending the metro with two extra routes were made in 1895, but the lines only opened more than 70 years later in 1970 and 1976. Following retirement in the early 1970s, car No.18 was preserved and is on display at the Seashore Trolley Museum in Kennebunkport, US.

◁ NER petrol-electric autocar, 1903

Wheel arrangement	2 x 4-wheel bogies (1 powered)
Transmission	2 traction motors
Engine	petrol
Total power output	80 hp (59.6 kW)
Top speed	approx. 36 mph (58 km/h)

Two of these petrol-electric railcars were built in 1903 in the UK at the North Eastern Railway's York Works. The original Wolsey four-cylinder engine that drove generators to power the two electric traction motors was replaced by a six-cylinder 225 hp (168 kW) engine in 1923. The railcars had been withdrawn by 1931. One is being restored at the Embsay & Bolton Abbey Steam Railway in Yorkshire.

▷ Drehstrom-Triebwagen, 1903

Wheel arrangement	2 x 6-wheel bogies, outer axles motorized
Power supply	6–14 kV DC (25–50 Hz)
Power rating	1,475 hp (1,100 kW)
Top speed	130 mph (210 km/h)

Built by Siemens & Halske and AEG of Germany and fitted with three-phase induction motors, two prototype high-speed Drehstrom-Triebwagen railcars were tested on the Prussian military railway south of Berlin in 1903. Taking overhead power from a triple catenary, the AEG-built railcar reached 130 mph (210 km/h) between Zossen and Marienfelde on 28 October 1903, a world rail-speed record not broken until 1931.

▷ NER electric locomotive, 1905

Wheel arrangement	Bo-Bo
Power supply	600–630 V DC, third-rail or catenary
Power rating	640 hp (477 kW)
Top speed	approx. 27 mph (43 km/h)

Drawing power from either a third-rail or an overhead catenary, two of these locomotives were built by British Thomson-Houston for the North Eastern Railway in 1903-04 but was not operational until 1905 when the line was electrified. They worked on a steeply graded freight line to a quayside in Newcastle-upon-Tyne until 1964. One is preserved at the Locomotion Museum in Shildon, County Durham.

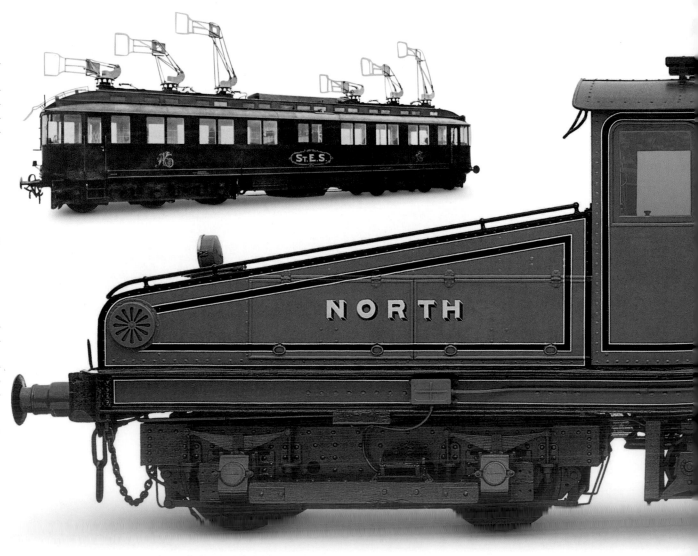

TALKING POINT

Ticketing on the Railways

Early railway companies issued tickets to passengers on handwritten pieces of paper. This was time-consuming and open to fraud by unscrupulous ticket clerks. Invented by Thomas Edmondson, an English station master, the Edmondson railway ticket system was introduced in 1842. Using preprinted, durable cards was not only a faster means of issuing tickets but they were also given unique serial numbers that had to be accounted for by booking clerks each day. Ticket inspectors at stations and on trains punched holes in the tickets to prevent reuse.

Ticket punch Featuring a decorative, three-pointed spike, this silver ticket punch was made by the Bonney-Vehslage Tool Co. for the Baltimore & Ohio Railroad in 1906.

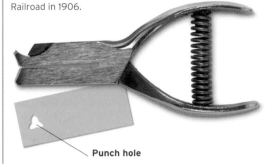

← **Punch hole**

△ **B&O Bo Switcher, 1895**

Wheel arrangement	Bo (0-4-0)
Power supply	approx. 450 V, catenary
Power rating	approx. 15 hp (11.2 kW)
Top speed	approx. 10 mph (16 km/h)

Opened in 1860, the Baltimore & Ohio Railroad's network of railways serving waterfront warehouses at Fells Point in Baltimore was originally horsedrawn. Overhead streetcar power lines were introduced in 1896 with small electric switchers, like this No. 10 built by General Electric in 1909, taking over from horsepower.

▷ **Schynige Platte Class He2/2, 1910**

Wheel arrangement	0-4-0
Power supply	1,500 V DC, overhead catenary
Power rating	295 hp (220 kW)
Top speed	approx. 5 mph (8 km/h)

The 2-ft 7½-in- (0.8-m-) gauge Schynige Platte Railway in the Swiss Bernese Oberland opened using steam power in 1893. This steeply graded mountain rack railway was electrified in 1914. Four of the original electric engines built by the Swiss Locomotive & Machine Works and Brown Boveri still operate on the railway.

1914-1939
STEAM'S
ZENITH

The New Empire State Express Passing West Point in the Highlands of the Hudson on the New York Central System

New York Central System

STEAM'S ZENITH

In 1914 the world was plunged into a terrible conflict. World War I ("the Great War") lasted until 1918, and during the four years of hostilities railways played a key role. The ability to move men, munitions, and supplies by train assumed new importance; in many countries full-size locomotives were specially built for the military, while narrow-gauge railways were created to serve the war effort. The latter were designed to be laid easily and to run close by the front lines.

△ **Ticket for the *Royal Blue*, 1935**
Recalling the glamour of the original *Royal Blue* train, B&O Railroad marketed the revamped service as elegant and luxurious.

At the end of the war, maps were redrawn, and many new or recreated countries found themselves inheriting existing rail systems, which they adapted to meet particular demands inside their new borders. In Germany, a post-war reorganization brought its railways together to create the Deutsche Reichsbahn. In Britain, the government merged the private rail companies to form what became known as the "Big Four".

A desire for progress and increasing rivalry (as well as competition from cars and aeroplanes) combined to give rise to a new age of speed and streamlining. As the Art Deco visual style took hold across the world, new, futuristic-looking trains were launched. Railways rivalled each other not only through offering greater speed and comfort, but also through clever marketing. Towards the end of the period, in July 1938, Britain's *Mallard* snatched the steam speed record from a German locomotive by reportedly reaching 126 mph (203 km/h) – a figure that officially has never been beaten.

Yet as steam neared its streamlined zenith, the push for speed and modernity created a new breed alongside the giants of steam and new lightweight diesel trains began to appear in North America and Europe during the 1930s.

> " There is **more poetry** in the rush of a **single railroad train** across the continent than in all **the gory story of Troy**"
>
> JOAQUIN MILLER, US POET

◁ **The new *Empire State Express* poster** by Leslie Ragan advertises the US's burgeoning railroad tourism in the 1930s

Key Events

▷ **1914** Outbreak of World War I. Railways prove to be essential for the transportation of troops and supplies.

▷ **1915** In Germany, Leipzig's main station is completed – the world's largest station measured by floor area.

▷ **1916** The final section of the Trans-Siberian Railway is opened.

▷ **1917** The Trans-Australian Railway is finished. Its route includes the world's longest stretch of straight track at nearly 300 miles (483 km).

▷ **1920** Germany's railways come under the new Deutsche Reichsbahn.

▷ **1931** Germany's petrol-powered Schienenzeppelin reaches 143 mph (230 km/h), setting a rail speed record.

▷ **1935** The first section of the Moscow Metro opens.

▷ **1934** Sir Nigel Gresley's steam locomotive *Flying Scotsman* records a speed of 100 mph (161 km/h).

▷ **1936** A German "Leipzig" diesel railcar travels at 127 mph (205 km/h), a record for diesel traction.

▷ **1938** France's railways are brought together as the Société Nationale des Chemins de fer Français (SNCF).

▷ **1938** Sir Nigel Gresley's *Mallard* hits 126 mph (203 km/h) – a steam speed record that stands today.

△ **Record-breaking *Mallard***
Mallard and the dynamometer car stand at Barkston on Sunday 3 July 1938, braced for the run that will earn the locomotive a world speed record for steam.

Locomotives for World War I

Following the outbreak of World War I, the Railway Operating Division (ROD) of the British Royal Engineers was formed in 1915 to operate railways in the European and Middle East theatres of war. The British network of narrow-gauge trench railways was operated by the War Department Light Railways, while the French had already standardized portable, 1-ft 11¾-in (0.60-m) gauge, military Decauville equipment to supply ammunition and stores to the Western Front. The Germans used a similar system for their trench railways – the Heeresfeldbahn. The entry of the US into the war in 1917 saw many US-built locomotives shipped across the Atlantic for service in France.

△ GWR Dean Goods, 1883

Wheel arrangement	0-6-0
Cylinders	2 (inside)
Boiler pressure	180 psi (12.65 kg/sq cm)
Driving wheel diameter	61¾ in (1,570 mm)
Top speed	approx. 45 mph (72 km/h)

Designed by William Dean, 260 of these standard-gauge freight locomotives were built at the Great Western Railway's Swindon Works between 1883 and 1899. In 1917 the Railway Operating Division commandeered 62 of them to operate supply trains in northern France. Some also served in France during WWII.

△ Baldwin Switcher, 1917

Wheel arrangement	0-6-0T
Cylinders	2
Boiler pressure	190 psi (13.4 kg/sq cm)
Driving wheel diameter	48 in (1,220 mm)
Top speed	approx. 30 mph (48 km/h)

Built in the US by the Baldwin Locomotive Works, the 651–700 Series of Railway Operating Division shunting (or switching) locomotives was introduced in 1917 for use by the British Military Railways in France. After the war they became Class 58 of the Belgian National Railways.

▽ Henschel metre-gauge, 1914

Wheel arrangement	0-6-0T
Cylinders	2
Boiler pressure	200 psi (14 kg/sq cm)
Driving wheel diameter	31½ in (800 mm)
Top speed	approx. 18 mph (29 km/h)

Built by the German company Henschel in 1914, two of these 3-ft 3-in- (1-m-) gauge locomotives were originally supplied to the Army Technical Research Institute. They were later transferred to the Nordhausen-Wernigerode Railway in the Harz Mountains in central Germany, where they hauled trains carrying standard-gauge freight wagons.

▷ O&K Feldbahn, 1903

Wheel arrangement	0-8-0T
Cylinders	2
Boiler pressure	approx. 180 psi (12.65 kg/sq cm)
Driving wheel diameter	approx. 22¾ in (580 mm)
Top speed	approx. 15 mph (24 km/h)

Introduced in 1903, around 2,500 of these 1-ft 11¾-in- (0.60-m-) gauge "Brigadelok" locomotives were built by several German companies, and widely used on the military light railways constructed to supply forward positions of the German army. The locomotive shown here is No. 7999, an Orenstein & Koppel engine built in 1915 with Klein-Linder articulation of the front and rear axles.

△ GCR Class 8K, 1911

Wheel arrangement	2-8-0
Cylinders	2
Boiler pressure	180 psi (12.65 kg/sq cm)
Driving wheel diameter	56 in (1,420 mm)
Top speed	approx. 45 mph (72 km/h)

The Great Central Railway's Class 8K freight locomotive introduced in 1911 was chosen as the standard British Railway Operating Division 2-8-0 locomotive during WWI. A total of 521 were built, with many seeing service hauling troop and freight trains in France. During WWII many of these locomotives were sent on active service to the Middle East

△ **Baldwin ALCO narrow-gauge, 1916**

Wheel arrangement	4-6-0PT
Cylinders	2
Boiler pressure	178 psi (12.51 kg/sq cm)
Driving wheel diameter	23$^{1}/_{4}$in (590 mm)
Top speed	approx. 18 mph (29 km/h)

Based on a French design, these 1-ft 11$^{3}/_{4}$-in-(0.60-m-) gauge pannier tank locomotives were supplied by the Baldwin Locomotive Works and the American Locomotive Co. in the US to the British War Office, for use on front-line military railways in northern France and the Middle East during WWI.

TECHNOLOGY

Armoured Engines

The British pioneered the use of small, armoured, narrow-gauge petrol locomotives to operate on the temporary railways that served the front line during World War I. Unlike steam locomotives, which could easily be spotted by the enemy, these locomotives could haul ammunition trains to forward positions during daylight hours without being detected.

Simplex locomotive Built for the British War Office by Motor Rail Ltd in 1917, this 1-ft 11$^{3}/_{4}$-in (0.60-m), four-wheel, engine hauled 15-ton (15.2-tonne) ammunition trains at 5 mph (8 km/h) to the trenches in northern France.

△ **Pershing Nord, 1917**

Wheel arrangement	2-8-0
Cylinders	2
Boiler pressure	189 psi (13.28 kg/sq cm)
Driving wheel diameter	56 in (1,420 mm)
Top speed	56 mph (90 km/h)

The North British Locomotive Co. in Glasgow supplied 113 Consolidation Pershings for the Compagnie des Chemins de fer du Nord in France. While the railway was happy to run these large locomotives at up to 56 mph (90 km/h), other French railways preferred lower operating speeds.

△ **Baldwin "Spider", 1917**

Wheel arrangement	4-6-0
Cylinders	2
Boiler pressure	190 psi (13.4 kg/sq cm)
Driving wheel diameter	61$^{3}/_{4}$in (1,570 mm)
Top speed	approx. 65 mph (105 km/h)

Nicknamed "Spiders" by British soldiers, 70 of these mixed-traffic locomotives were built with bar frames by the US Baldwin Locomotive Works between 1917 and 1918 for service on the Western Front during WWI. Later they became Class 40 of the Belgian National Railways.

War Machines

Railway-mounted artillery featured in conflicts from the American Civil War to World War II. They caused great destruction, most notably during World War I when Germany's *Pariskanonen* (Paris Guns) bombarded the French capital with 230-lb (106-kg) shells from a distance of 75 miles (120 km). Rail-mounted gun turrets provided fast, mobile fire power, and being moveable, the guns could also be hidden from enemy attacks. From 1862 to 1945, before the introduction of air attacks, they were perhaps the most destructive long-range weapons.

ON THE FRONT LINE

Austria-Hungary, Britain, France, Germany, Russia, and the US all deployed rail-mounted artillery. While France was the first nation to equip its army with rail-mounted howitzers by adapting naval guns for use on railway wagons, Germany developed howitzers that set records for size, range, and destructive power. Companies such as Krupp and Skoda built *Dicke Bertha* (Big Bertha) and *Schlanke Emma* (Skinny Emma) howitzers respectively, both of which inflicted great damage on French and Belgian defences. However, Krupp's *Schwerer Gustav* (Heavy Gustavs) – the biggest land weapon ever – was a failure. Though capable of firing 4-ton (4,064-kg) shells as far as 29 miles (46 km), it demanded a crew of no fewer than 1,420 people and two parallel tracks, which made the gun so impractical that it only saw action once.

French railway guns were engaged in the Somme offensive during World War I. The weapon pictured required a crew of 15 men.

Fast and Powerful

The introduction of longer and heavier express passenger trains in Europe and the US during the 1920s and 1930s led to the building of more powerful and faster types of locomotives to standard designs. In Britain, Sir Nigel Gresley led the way with his three-cylinder A1 and A3 Pacific 4-6-2s of which *Flying Scotsman* is justifiably world famous. Other British locomotive engineers such as the Great Western Railway's Charles Collett and the London, Midland & Scottish Railway's Henry Fowler favoured a 4-6-0 wheel arrangement. In the US, Germany, and France, the Pacific type became the favoured express passenger locomotive type.

△ PRR Class K4s, 1914

Wheel arrangement	4-6-2
Cylinders	2
Boiler pressure	205 psi (14.4 kg/sq cm)
Driving wheel diameter	80 in (2,030 mm)
Top speed	approx. 70 mph (113 km/h)

The Class K4s Pacific locomotives, of which 425 were built in the US between 1914 and 1928, were the Pennsylvania Railroad's premier express steam locomotive. They were often used in double or triple headers to haul heavy trains.

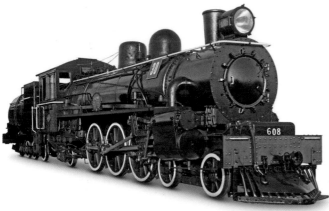

◁ NZR Class Ab, 1915

Wheel arrangement	4-6-2
Cylinders	2
Boiler pressure	180 psi (12.65 kg/sq cm)
Driving wheel diameter	54 in (1,372 mm)
Top speed	approx. 60 mph (96 km/h)

One of a class of 141 locomotives, New Zealand Railways Class Ab Pacific locomotive No. 608 is named *Passchendaele* in memory of NZR staff killed in WWI. Ab engines were replaced by diesels in the 1960s but five have been preserved.

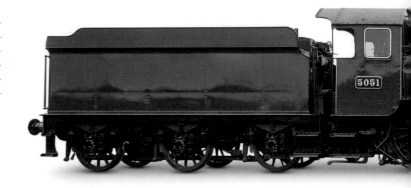

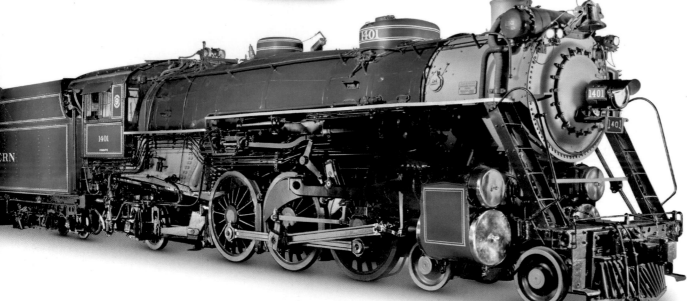

◁ SOU Class Ps-4, 1923

Wheel arrangement	4-6-2
Cylinders	2
Boiler pressure	200 psi (14.06 kg/sq cm)
Driving wheel diameter	73 in (1,854 mm)
Top speed	approx. 80 mph (129 km/h)

Finished in a striking green livery, the 64 Class Ps-4 Pacific-type express passenger locomotives were built for the Southern Railway of the US by the American Locomotive Company (ALCO) and the Baldwin Locomotive Works between 1923 and 1928. Designed to haul the railroad's heavy expresses, they had been replaced by diesels by the early 1950s. No. 1401 is on display in the Smithsonian Institution in Washington DC.

△ PRR Class G5s, 1924

Wheel arrangement	4-6-0
Cylinders	2
Boiler pressure	205 psi (14.4 kg/sq cm)
Driving wheel diameter	68 in (1,730 mm)
Top speed	approx. 70 mph (113 km/h)

This engine was designed by William Kiesel to work commuter trains on the Pennsylvania Railroad. The Class G5s was one of the largest and most powerful 4-6-0s in the world. No. 5741 is on display in the Railroad Museum of Pennsylvania.

▷ LMS Royal Scot Class, 1927

Wheel arrangement	4-6-0
Cylinders	3
Boiler pressure	250 psi (17.57 kg/sq cm)
Driving wheel diameter	81 in (2,057 mm)
Top speed	approx. 80 mph (129 km/h)

Designed by Sir Henry Fowler, 70 Royal Scot Class locomotives were built to haul long distance express trains on the London, Midland & Scottish Railway. They were later rebuilt by William Stanier with Type 2A tapered boilers, and remained in service until the early 1960s.

BUILDING THE 'ROYAL SCOT' ENGINE

◁ DR Class 01, 1926

Wheel arrangement 4-6-2

Cylinders 2

Boiler pressure 232 psi (16.3 kg/sq cm)

Driving wheel diameter 78³/₄in (2,000 mm)

Top speed approx. 81 mph (130 km/h)

A total of 241 (including 10 rebuilt Class 02s) of these standardized Class 01 express locomotives were built for the Deutsche Reichsbahn between 1926 and 1938. Some engines remained in service in East Germany until the early 1980s.

△ LNER Class A3, 1928

Wheel arrangement 4-6-2

Cylinders 3

Boiler pressure 220 psi (15.46 kg/sq cm)

Driving wheel diameter 80 in (2,030 mm)

Top speed 108 mph (174 km/h)

Britain's Sir Nigel Gresley designed the A3 for the London & North Eastern Railway. These locomotives hauled express trains between London's King's Cross and Scotland. No. 4472 *Flying Scotsman* is the only example preserved.

△ GWR Castle Class, 1936

Wheel arrangement 4-6-0

Cylinders 4

Boiler pressure 225 psi (15.82 kg/sq cm)

Driving wheel diameter 80¹/₂in (2,045 mm)

Top speed approx. 100 mph (161 km/h)

These express locomotives were designed by Charles Collett for the Great Western Railway. Its Swindon Works built 171 Castle Class engines between 1923 and 1950. Shown here is No. 5051. They had all been retired by 1965, but eight have now been preserved. No. 5051 *Drysllyn Castle* is at Didcot Railway Centre.

△ GWR King Class, 1930

Wheel arrangement 4-6-0

Cylinders 4

Boiler pressure 250 psi (17.57 kg/sq cm)

Driving wheel diameter 78 in (1,980 mm)

Top speed approx. 90 mph (145 km/h)

The King Class was designed by Charles Collett for the Great Western Railway. Thirty of these express locomotives were built at Swindon Works in England between 1927 and 1936. They were replaced by diesels in the early 1960s; three including this one, No. 6023 *King Edward II*, have been preserved.

▷ Nord Pacific, 1936

Wheel arrangement 4-6-2

Cylinders 4 (compound)

Boiler pressure 240 psi (16.87 kg/sq cm)

Driving wheel diameter 75¹/₂in (1,918 mm)

Top speed approx. 81 mph (130 km/h)

French engineer André Chapelon designed these powerful locomotives for the Compagnie du Nord. They hauled express trains such as the *Flèche d'Or* in northern France. Shown here is No. 3.1192, which is exhibited at the Cité du Train, Mulhouse, France.

King Edward II

Built at Swindon Works in June 1930, the King Class locomotive No. 6023 *King Edward II* was in a class of engines considered to be the most powerful machines on any British railway. The first of the class, No. 6000, built in 1927, was named after the reigning monarch - King George V; later engines carried names of earlier kings in reverse order of ascendance. *King Edward II* served for 32 years, first with the Great Western Railway then British Railways.

DESIGNED BY CHARLES B. COLLETT, the King Class was a natural progression from his four-cylinder 4-6-0 Castle Class engines, which had enjoyed great success. Commentators at the time even wondered whether this was a new design or simply a "super" Castle. The King Class locomotives were able to handle the heaviest trains operated by the GWR, but their heavy axle weight restricted them to the London–Plymouth and London–Wolverhampton (via Bicester) routes. Owing to this limited route availability, relatively few were built.

After being withdrawn from service in June 1962, No. 6023 *King Edward II* was sold to locomotive scrap merchants Woodham Brothers of Barry, South Wales and remained there until its rescue in December 1984. By this time the engine was a rotting hulk and its rear driving wheel set had been sliced through by a cutting torch following a shunting mishap. This iconic locomotive has now been fully restored and it returned to steam at Didcot Railway Centre in 2011.

FRONT VIEW

REAR VIEW

British Railways logo
The original lion emblem, known as the "Cycling Lion", was used on locomotives between 1950 and 1956.

SPECIFICATIONS			
Class	King	In-service period	1930-62 (*King Edward II*)
Wheel arrangement	4-6-0	Cylinders	4
Origin	UK	Boiler pressure	250 psi (17.5 kg/sq cm)
Designer/builder	CB Collett/Swindon Works	Driving wheel diameter	78 in (1,980 mm)
Number produced	30 King Class	Top speed	approx. 110 mph (177 km/h)

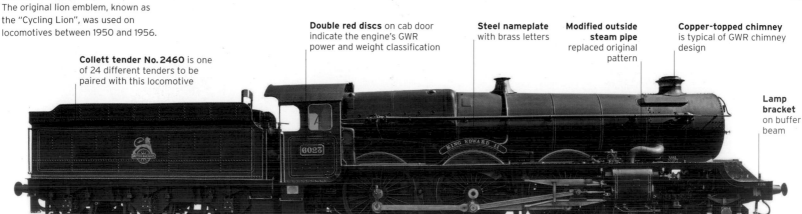

Collett tender No. 2460 is one of 24 different tenders to be paired with this locomotive

Double red discs on cab door indicate the engine's GWR power and weight classification

Steel nameplate with brass letters

Modified outside steam pipe replaced original pattern

Copper-topped chimney is typical of GWR chimney design

Lamp bracket on buffer beam

Long service history
King Edward II is one of only three surviving members of its class and performed over 1.5 million miles (2.414 million km) of service. The "PDN" mark on the buffer framing is the GWR code for the Old Oak Common depot where it was first based.

6023

PDN

EXTERIOR

The King Class engines were originally turned out in the GWR's traditional green Swindon livery, with their distinctive, copper-topped chimney. However, in 1948 two King Class locomotives were turned out in an experimental dark blue livery with red, cream, and grey lining. In 1950 a standard Caledonian blue livery with black and white lining was introduced. Over time British Railways changed the livery back to green. *King Edward II* has been restored in the BR 1950s blue livery.

1. Nameplate in brass letters on steel   2. Numberplate on cab side   3. Interior of smokebox   4. Chimney with polished copper cap   5. Axle and leaf spring suspension on front set of bogie wheels   6. Retaining valve for vacuum brake changeover, and copper pipes for lubricator   7. Crosshead of inside cylinder, seen through inspection hole   8. Copper pipes for directing steam from cylinder cocks   9. Crosshead and slidebars   10. Big end bearing of connecting rod   11. Cladding sheets on side of outer firebox   12. Vacuum brake ejector   13. Builder's plate on rear of tender tank   14. Speedometer drive   15. Low-level tender filler (a modern addition)   16. Buffer at rear of tender   17. Front of tender viewed from cab

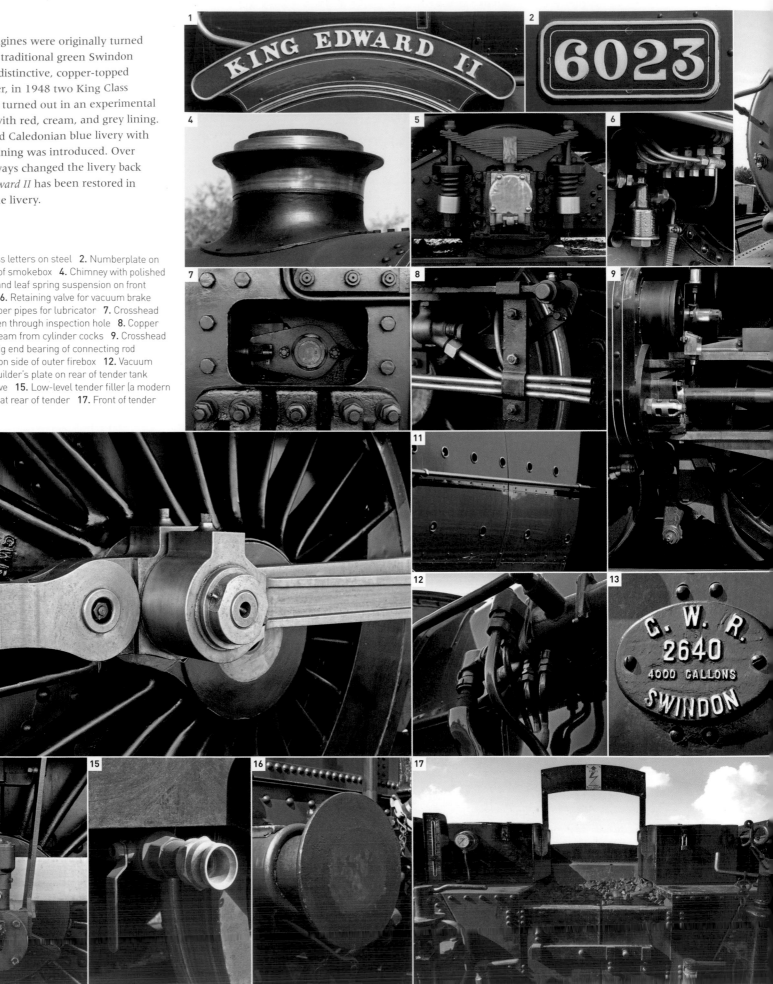

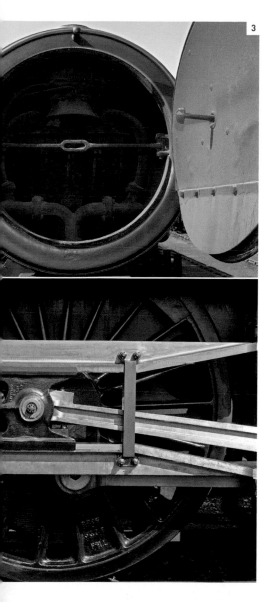

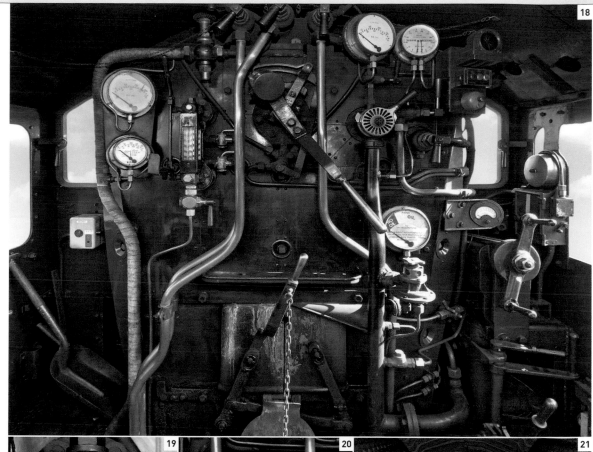

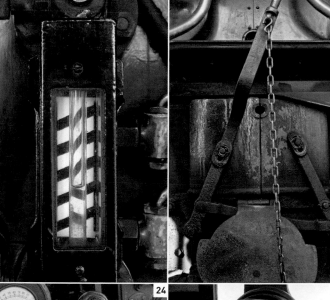

CAB INTERIOR

The King Class footplate layout followed the Swindon Works' standard design, which was practical and reasonably spacious. Early locomotives were generally of the right-hand drive configuration, with the fireman's seat being on the left or nearside. When double-track railways first came into being the lineside signals were placed on the near side, so many railway companies changed their footplate designs to left-hand drive. However, the GWR continued to configure their locomotives for right-hand driving. Unlike other designers Collett did not include padded seating for the footplate crew, preferring instead a simple, hinged wooden seat.

18. Cab controls on backhead of firebox
19. Water-level gauge **20.** Firehole door
21. Interior of firebox **22.** Vacuum brake control
23. Mechanical lubricator gauge **24.** Screw
reverser (clockwise forwards, anticlockwise
backwards) **25.** Automatic Train Control (ATC)
audible signalling system **26.** Wooden seat on
fireman's side of cab

Great Journeys
Orient Express

Made famous in literature and film, the *Orient Express* was the brainchild of the Belgian Georges Nagelmackers, founder of the Compagnie Internationale des Wagons-Lits, a company that specialized in operating luxury train services on European railways.

FOLLOWING A SUCCESSFUL TEST JOURNEY
between Paris and Vienna in 1882, the first regular *Express d'Orient* left Gare de l'Est, Paris behind an outside cylinder Est 2-4-0 locomotive on 4 October 1883. It travelled eastwards to Strasbourg and then to Munich before crossing into Austria and calling at Salzburg and Vienna. From here the train continued on to Budapest, Bucharest, and Giurgiu on the banks of the River Danube in Romania. Passengers were then ferried across the river to Rustchuk in Bulgaria, where they boarded older rolling stock of the Austrian Eastern Railway to Varna on the Black Sea coast. From Varna, passengers then made an 18-hour sea voyage to Constantinople. Between Paris and Giurgiu the train consisted of five new bogie sleeping cars, a bogie restaurant car, and two baggage cars, all built to a high standard in teak, with locomotives changed many times *en route*. The journey took four days in total so passengers had plenty of time to enjoy the high standard of cuisine in the restaurant car on the first leg of the journey.

From 1889 the train began running directly between Paris and Constantinople, following the opening of new railways through the Balkans in Serbia, Bulgaria, and European Turkey. The train was renamed the *Orient Express* in 1891.

Services ended with the onset of World War I, but recommenced after the war and the train once again became popular. The Simplon Tunnel had opened under the Alps in 1906, and in 1919 a new *Simplon-Orient-Express* took a route between

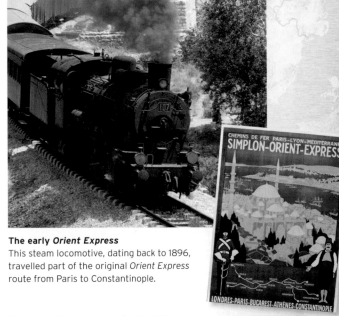

The early *Orient Express*
This steam locomotive, dating back to 1896, travelled part of the original *Orient Express* route from Paris to Constantinople.

VINTAGE POSTER FROM 1920S

Paris and Constantinople via Milan, Venice, Trieste, and Belgrade. By the 1930s three separate trains operated: the *Orient Express* on the original 1889 route; the *Simplon-Orient-Express* via the Simplon Tunnel; and the *Arlberg-Orient-Express* via Zurich, Innsbruck, and Budapest with through carriages for Athens. Sleeping cars started running from Calais, providing the first transcontinental journey across Europe.

Following suspension during World War II, the *Orient Express* resumed service in 1952, but both it and the *Arlberg-Orient-Express* had ceased to run by 1962. The *Simplon-Orient-Express* was replaced that year by the *Direct-Orient-Express*, which was withdrawn in 1977. Some of the carriages were bought by a private company in 1982, which now runs *Venice Simplon-Orient Express* services to several destinations in Europe.

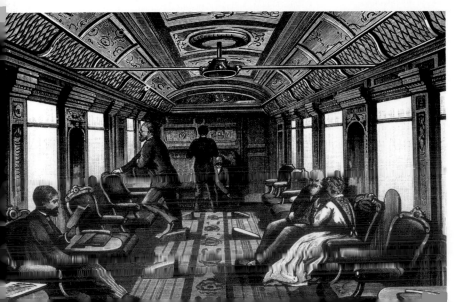

Splendour on the Orient Express
The saloon car aboard the *Orient Express* around 1896 was designed for luxury, featuring detailed wood panelling and an inlaid ceiling.

UNITED KINGDOM

NORTH SEA

London

ENGLISH CHANNEL

• Calais

Black Forest **1**
The *Orient Express* passed through the Black Forest wooded mountain range in Baden in southwestern Germany.

Paris •

Strasbourg

Zurich •

SWITZERLAND

Milan •

FRANCE

Monaco •

Last journey The last run of the *Direct-Orient-Express* terminated off-route in Monaco in 1977. The 1920s carriages were purchased at auction here and later restored, primarily for use on the London to Venice route.

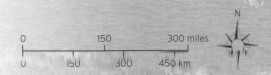

| 0 | 150 | 300 miles |
| 0 | 150 | 300 | 450 km |

N

KEY FACTS

DATES

1883 First regular *Express d'Orient* leaves Gare de l'Est, Paris for Giurgiu in Romania
1889 First through service Paris-Constantinople
1891 Train renamed *Orient Express*
1977 Regular Paris-Istanbul journeys cease

TRAIN

Locomotive In France the first *Orient Express* was hauled by a Chemins de Fer de l'Est outside cylinder 2-4-0. Many different locomotives were used
Carriages (1883) 5 bogie sleeping cars with accommodation for 20 passengers and 2 washrooms; 1 bogie restaurant car; 1 baggage car; 1 mail car

JOURNEY

1883 Paris to Constantinople (original journey)
Train from Paris to Giurgiu; passengers ferried from Giurgiu across Danube to Rustchuk, Bulgaria followed by a 7-hour train to Varna; ship to Constantinople (Istanbul); approx. 1,500 miles (2,414 km), 4 days
1889 Paris to Constantinople
Train diverted at Budapest to Belgrade and Nis, Serbia, through Dragoman Pass in Passara Mountains to Bulgaria, Pazarzhik to Plovdiv, then Constantinople; approx. 1,400 miles (2,250 km), 67 hours 35 minutes
Paris to Istanbul (current, not shown on map)
Runs annually; approx. 1,400 miles (2,253 km), 6 days, 5 nights; spends the night in Budapest and Bucharest
London to Venice (current VSOE route, not on map)
Route via Paris/Innsbruck/Verona; 1,065 miles (1,714 km), 2 days, 1 night

RAILWAY

Gauge Standard 4ft 8½in (1.435 m)
Tunnels Longest (*Simplon-Orient Express* route) is Simplon Tunnel, Alps 65,039 ft (19,824 m)
Highest point Simplon Tunnel, Alps 2,313 ft (705 m)

KEY
● Start/Finish
● Main stations
▦ Original route
▦ 1889 route
━ Change of train
▪▪▪ Sea voyage

A ROMANTIC ADVENTURE

The original route of the *Orient Express* is now retraced annually, with all the luxury of the earliest trips. The antique train passes through seven countries, with numerous stop-offs along the way. Plush private cabins, personal stewards, and gourmet meals can all be expected on board.

2 Through the Alps
The train passed through the ski resort of St Anton in the Tyrolean Alps in Austria along the route.

3 Budapest
Until unification in 1873, Budapest, the capital of Hungary, was two separate cities: Buda was on the west bank of the River Danube while Pest was on the east bank.

5 Varna
The ancient port city of Varna dates back five millennia and its necropolis is the site of the oldest find of gold metallurgy.

Giurgiu Passengers were ferried across the Danube from Giurgiu to board another train at Rustchuk.

Iron Gates 4
This gorge on the Danube forms the boundary between Serbia and Romania.

Sirkeci Station 6
The Orient Express Restaurant is located at the historic Sirkeci Station at Istanbul, the original route's terminus.

GERMANY

Munich
Innsbruck
Salzburg
AUSTRIA
Vienna
HUNGARY
Budapest
Verona
Trieste
Venice
ITALY
Belgrade
SERBIA
Nis
ROMANIA
Bucharest
Giurgiu
Rustchuk (Ruse)
Constanta
Varna
BULGARIA
Plovdiv
Pazardzhik
BLACK SEA
Constantinople (Istanbul)
TURKEY
Athens

Mixed-traffic Movers

By the 1930s the standardization of machine parts by European and US locomotive builders had reduced construction and maintenance costs significantly. Powerful engines designed to haul express freight and passenger trains were soon coming off the production lines in great numbers. In Britain both Charles Collett of the Great Western Railway (GWR) and William Stanier of the London, Midland & Scottish Railway (LMS) made standardization a common theme when designing their new 4-6-0 locomotives, while in Germany the Class 41 2-8-2s built for the Deutsche Reichsbahn incorporated parts simultaneously developed for three other classes.

△ LMS Class 5MT, 1934

Wheel arrangement	4-6-0
Cylinders	2
Boiler pressure	225 psi (15.82 kg/sq cm)
Driving wheel diameter	72 in (1,830 mm)
Top speed	approx. 80 mph (129 km/h)

Designed by William Stanier for the London, Midland & Scottish Railway, many of these powerful mixed-traffic locomotives, "Black Fives", saw service in Britain until the end of steam in 1968. A total of 842 were built.

◁ SR S15 Class, 1927

Wheel arrangement	4-6-0
Cylinders	2
Boiler pressure	175–200 psi (12.30–14 kg/sq cm)
Driving wheel diameter	67 in (1,700 mm)
Top speed	approx. 65 mph (105 km/h)

These powerful British locomotives were a modified version of an earlier Robert Urie design, introduced by Richard Maunsell. They were built by the Southern Railway at its Eastleigh Works in Southern England.

◁ NZR Class K, 1932

Wheel arrangement	4-8-4
Cylinders	2
Boiler pressure	200 psi (14.06 kg/sq cm)
Driving wheel diameter	54 in (1,372 mm)
Top speed	approx. 65 mph (105 km/h)

Built to haul heavy freight and passenger trains on New Zealand's mountainous North Island, 30 of the Class Ks were built at Hutt Workshops for New Zealand Railways between 1932 and 1936. They were gradually withdrawn from service between 1964 and 1967.

△ LNER Class V2, 1936

Wheel arrangement	2-6-2
Cylinders	3
Boiler pressure	220 psi (15.46 kg/sq cm)
Driving wheel diameter	74 in (1,880 mm)
Top speed	approx. 100 mph (161 km/h)

These engines were designed by Sir Nigel Gresley for the London & North Eastern Railway and hauled both express passenger and express freight trains. No. 4771 *Green Arrow* is the only preserved example.

△ DR Class 41, 1937

Wheel arrangement	2-8-2
Cylinders	2
Boiler pressure	290 psi/228 psi (20.39 kg/sq cm/16 kg/sq cm)
Driving wheel diameter	63 in (1,600 mm)
Top speed	approx. 56 mph (90 km/h)

Built with parts that were designed for several different locomotive types, these powerful, fast freight engines were constructed for the Deutsche Reichsbahn between 1937 and 1941.

TALKING POINT

Fresh Milk

Transporting perishable goods such as milk, fish, and meat by rail called for specialized freight wagons. In Britain, milk was first conveyed in milk churns loaded into ventilated wagons at country stations, but from the 1930s it was carried in six-wheeled milk tank wagons loaded at a creamery. The wagons were marshalled into trains and hauled by powerful express steam locomotives to depots in and around London. The last milk trains to operate in Britain ran in 1981.

London's dairy supplier With a capacity of 3,000 gallons (13,638 litres), the Express Dairy six-wheel milk tank wagon weighed as much as a loaded passenger coach when full. This wagon was built by the Southern Railway in 1931 and rebuilt in 1937.

◁ GWR Manor Class, 1938

Wheel arrangement	4-6-0
Cylinders	2
Boiler pressure	225 psi (15.82 kg/sq cm)
Driving wheel diameter	68 in (1,730 mm)
Top speed	approx. 65 mph (105 km/h)

With their light axle loading, these Great Western Railway mixed-traffic locomotives could operate on secondary and branch lines as well as main lines in England and Wales. This engine is No. 7808 *Cookham Manor*.

◁ GWR Hall Class, 1928

Wheel arrangement	4-6-0
Cylinders	2
Boiler pressure	225 psi (15.82 kg/sq cm)
Driving wheel diameter	72 in (1,830 mm)
Top speed	approx. 70 mph (113 km/h)

A total of 259 of these versatile engines, designed by Charles Collett, were built at the Great Western Railway's Swindon Works between 1928 and 1943. This is No. 5900 *Hinderton Hall*.

Versatile Engines

While the development of more powerful and faster express steam locomotives gathered pace during the 1920s and 1930s there was also the parallel development of smaller engines designed for shunting (or switching) at freight yards, railway workshops, and stations, or to carry out passenger and freight duties on country branch lines. Many of these versatile locomotives remained in active service until the end of the steam era, while some have since been restored to service on heritage railways.

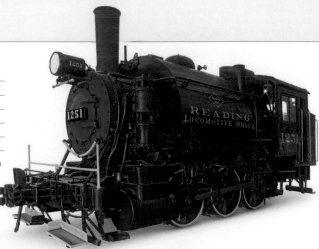

▷ P&R Switcher No. 1251, 1918

Wheel arrangement	0-6-0T
Cylinders	2
Boiler pressure	150 psi (10.53 kg/sq cm)
Driving wheel diameter	50 in (1,270 mm)
Top speed	approx. 25 mph (40 km/h)

Rebuilt in 1918 from a Class 1-2a Consolidation locomotive, No. 1251 spent its life as a switcher at the Philadelphia & Reading Railroad Shops in Reading, Pennsylvania. It was retired in 1964 as the last steam engine on a US Class 1 railroad, and is now on display at the Railroad Museum of Pennsylvania.

△ LMS Class 3F "Jinty", 1924

Wheel arrangement	0-6-0T
Cylinders	2 (inside)
Boiler pressure	160 psi (11.25 kg/sq cm)
Driving wheel diameter	55 in (1,400 mm)
Top speed	approx. 40 mph (64 km/h)

These tank locomotives, nicknamed "Jintys", were designed by Henry Fowler for the London, Midland & Scottish Railway. Widely used for shunting and local freight work in the Midlands and northwest England, 422 were built with the last examples remaining in service until 1967.

△ GWR 5600 Class, 1924

Wheel arrangement	0-6-2T
Cylinders	2 (inside)
Boiler pressure	200 psi (14.06 kg/sq cm)
Driving wheel diameter	55½ in (1,410 mm)
Top speed	approx. 45 mph (72 km/h)

Designed for the Great Western Railway by Charles Collett, 150 of these powerful tank engines were built at the company's Swindon Works and 50 by Armstrong Whitworth in Newcastle-upon-Tyne. They mainly saw service in the South Wales valleys hauling coal trains, but were also used on local passenger services.

▽ L&B *Lew*, 1925

Wheel arrangement	2-6-2T
Cylinders	2
Boiler pressure	160 psi (11.25 kg/sq cm)
Driving wheel diameter	33 in (840 mm)
Top speed	approx. 25 mph (40 km/h)

Completed at the Ffestiniog Railway's Boston Lodge Works in 2010, *Lyd* (shown) is a replica of *Lew*, which was built by Manning Wardle in 1925 for the Southern Railway's 1-ft 11¾-in (0.60-m) gauge Lynton to Barnstaple line. The line, closed in 1935, is now in the process of being reopened by enthusiasts.

TECHNOLOGY

Battery Power

Battery locomotives are powered by huge onboard batteries that are recharged in between duties. These engines were once used on railways serving industrial complexes, such as explosives and chemical factories, mines, or anywhere else where normal steam or diesel locomotives could present hazards, such as fire risk, explosion, or fumes. In England, the London Underground uses battery-electric locomotives when the normal electric power is turned off during periods of night-time maintenance.

English Electric EE788 0-4-0 Battery Locomotive This four-wheel, 70-hp (52-kW), battery-electric locomotive was built by English Electric at their Preston factory in England in 1930 and worked for many years at their Stafford Works. It is currently on display at the Ribble Steam Railway Museum in Preston.

**△ GWR 4575 Class
Prairie Tank, 1927**

Wheel arrangement	2-6-2T
Cylinders	2
Boiler pressure	200 psi (14.06 kg/sq cm)
Driving wheel diameter	55½ in (1,410 mm)
Top speed	approx. 50 mph (80 km/h)

Designed by Charles Collett, the 4575 Class of Prairie tank was built at the Great Western Railway's Swindon Works between 1927 and 1929. Of the 100 built, many saw service on branch line passenger and freight duties in England's West Country. No. 5572 shown here was one of the six fitted for push-pull operations. It is preserved at Didcot Railway Centre.

**△ GWR 5700 Class
Pannier Tank, 1929**

Wheel arrangement	0-6-0PT
Cylinders	2 (inside)
Boiler pressure	200 psi (14.06 kg/sq cm)
Driving wheel diameter	55½ in (1,410 mm)
Top speed	approx. 40 mph (64 km/h)

One of the most numerous classes of British steam engine, 863 of these Pannier Tanks were built for the Great Western Railway and British Railways between 1929 and 1950. They were usually seen at work on shunting duties or hauling passenger and freight trains on branch lines. Of the 16 preserved, No. 3738, seen here, is on display at Didcot Railway Centre.

△ DR Class 99.73-76, 1928

Wheel arrangement	2-10-2T
Cylinders	2
Boiler pressure	200 psi (14.06 kg/sq cm)
Driving wheel diameter	31½ in (800 mm)
Top speed	approx. 19 mph (31 km/h)

The Deutsche Reichsbahn had these tank engines built as a new standard design for 2-ft 5½-in- (0.75-m-) gauge lines in Saxony, eastern Germany. A number of these and a modified version introduced in 1950s are still in service today.

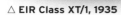

△ EIR Class XT/1, 1935

Wheel arrangement	0-4-2T
Cylinders	2
Boiler pressure	160 psi (11.25 kg/sq cm)
Driving wheel diameter	57 in (1,448 mm)
Top speed	approx. 40 mph (64 km/h)

Built by Freidrich Krupp AG of Berlin, Germany, for the 5-ft 6-in- (1.67-m-) gauge East Indian Railway, these locomotives were first introduced in 1929 and were used for light passenger work. No. 36863 (shown) was built in 1935 and is on static display at the National Rail Museum, New Delhi.

Freight Shifters

As train speeds rose, they increasingly carried a variety of goods, including perishable food items. Freight locomotives evolved accordingly. Mainland Europe and North America discarded the six-wheeler for front-rank duties, but the UK continued to build them. The 2-8-0, and variants on the eight-coupled wheelbase, became the main types. Canada, China, Germany, and the USSR built 10-coupled designs, but, especially in the US, the loads and terrain demanded nothing short of the giants.

◁ PRR Class A5s, 1917

Wheel arrangement	0-4-0
Cylinders	2
Boiler pressure	185 psi (13 kg/sq cm)
Driving wheel diameter	50 in (1,270 mm)
Top speed	approx. 25 mph (40 km/h)

The Pennsylvania Railroad served many industrial sites around Baltimore, Philadelphia, and New York, where a short-wheelbase switcher, or shunter, was essential to negotiate the tight clearances. One of the most powerful 0-4-0s ever, 47 of the Class A5s were built at the railroad's workshops in Altoona, Pennsylvania, up to 1924.

△ XE Class, 1928/30

Wheel arrangement	2-8-2
Cylinders	2
Boiler pressure	210 psi (14.8 kg/sq cm)
Driving wheel diameter	61½ in (1,562 mm)
Top speed	approx. 30 mph (48 km/h)

Aside from articulated types, the XE (X Eagle) Class of British-built Mikados (2-8-2s) were the largest steam locomotives on the subcontinent. A total of 93 of these broad-gauge (5-ft 6-in/1.67-m) designs were built, of which 35 were based in Pakistan after partition. No. 3634 *Angadh* is shown here.

▷ CP T1-C Class Selkirk, 1929

Wheel arrangement	2-10-4
Cylinders	2
Boiler pressure	285 psi (20.03 kg/sq cm)
Driving wheel diameter	63 in (1,600 mm)
Top speed	approx. 65 mph (105 km/h)

This semi-streamlined class of engines was built by Canadian Pacific Railway to master the Selkirk Mountains. Thirty of these oil-burners were built up to 1949, and were the largest and most powerful, non-articulated locomotives in the British Commonwealth. They hauled trains 262 miles (422 km) over the mountains from Calgary, Alberta, to Revelstoke, British Columbia.

▷ DR Class 44, 1930

Wheel arrangement	2-10-0
Cylinders	3
Boiler pressure	228 psi (16 kg/sq cm)
Driving wheel diameter	55 in (1,400 mm)
Top speed	approx. 50 mph (80 km/h)

The Deutsche Reichsbahn acquired the first 10 in 1926, but delayed further orders until 1937, after which no fewer than 1,979 were built up to 1949. Unusually for a freight design they had three cylinders, helping them to haul trains of up to 1,181 tons (1,200 tonnes).

Goods Wagons

By the 20th century, railways hauled loads ranging from salt to sugar, petrol to milk, and cattle to coal. Wagons evolved to cater for specific roles: hoppers transported coal, ores, and stone; tankers carried liquids and gases; and refrigerated cars carried perishable goods. Whatever the load, before the introduction of continuous braking, every train had a brake van. From here the guard kept watch over the train, using his brake to keep control of the loose-coupled wagons on down gradients and when stopping.

△ GWR "Toad" brake van, 1924

Type	Brake van
Weight	20 tons (20.32 tonnes)
Construction	wooden body on 4-wheel steel chassis
Railway	Great Western Railway

At a time when most UK goods trains lacked any form of through braking, the role of the guard was critical in controlling the train. From 1894 the Great Western Railway's guards manned "Toads", the name deriving from the

◁ **UP Challenger CSA-1 Class/
CSA-2 Class, 1936**

Wheel arrangement	4-6-6-4
Cylinders	4
Boiler pressure	280 psi (19.68 kg/sq cm)
Driving wheel diameter	69 in (1,753 mm)
Top speed	approx. 70 mph (113 km/h)

Union Pacific Railroad's Challenger proved that a simple articulated engine could haul huge loads at high speed. Each set of driving wheels was powered by two cylinders, with four trailing wheels to support the huge firebox. The American Locomotive Co. (ALCO) built 105 from 1936 to 1944. Two have been preserved, No. 3977 and No. 3985.

▷ **SAR Class 15F, 1938**

Wheel arrangement	4-8-2
Cylinders	2
Boiler pressure	210 psi (14.8 kg/sq cm)
Driving wheel diameter	60 in (1,524 mm)
Top speed	approx. 60 mph (96 km/h)

Most numerous of South African Railway's classes, the 15F was used predominantly in the Orange Free State and Western Transvaal. Construction spanned WWII; 205 were built by UK companies and a further 50 by German. Several have survived. The 1945-built No. 3007 is in the city of its birth at Glasgow's Riverside Museum.

◁ **GWR 2884 Class, 1938**

Wheel arrangement	2-8-0
Cylinders	2
Boiler pressure	225 psi (15.81 kg/sq cm)
Driving wheel diameter	55½ in (1,410 mm)
Top speed	approx. 45 mph (72 km/h)

The Great Western Railway's 2800 Class of 1903 – the first British 2-8-0 – was a success, persuading the GWR to add to the original total of 83. Modifications, though minor, merited a new designation – the 2884 Class, 81 of which were built from 1938 to 1942. No. 3822 is one of nine preserved.

◁ **FGEX fruit boxcar, 1928**

Type	Express refrigerated boxcar
Weight	24.73 tons (25.13 tonnes)
Construction	wooden body with integral cooling system mounted on steel underframe with two 4-wheel bogies
Railway	Fruit Growers' Express

A leasing company jointly owned by 11 railroads in the eastern and southeastern US, the Fruit Growers' Express built and operated several thousand refrigerated vehicles. Retired in the late 1970s, No. 57708 was preserved by the Cooperstown & Marne Railroad.

△ **ACF three-dome tanker, 1939**

Type	Three-dome bogie oil tanker
Weight	18.08 tons (18.37 tonnes)
Construction	steel superstructure mounted on a double bogie steel chassis
Railway	Shippers' Car Line Corporation

The American Car & Foundry Co. remains one of the major rolling stock manufacturers in the US. It built three-dome tanker No. 4556 in 1939 for the Shippers' Car Line Corporation. Riding on two four-wheel bogies, and used for transporting propane and liquid petroleum gas, the tanker has a capacity of 3,790 gallons (17,230 litres).

Herbert Nigel Gresley 1876–1941

Nigel Gresley's engineering career started at the age of 17, when he became an apprentice at Crewe Locomotive Works. After serving his apprenticeship he broadened his experience in the field by working as a fitter, designer, and tester, as well as the foreman of a running shed. In 1905 he began working for the Great Northern Railway where he designed locomotives, pioneered articulated carriages, and eventually rose to become Locomotive Superintendent in 1911. After the formation of the London & North Eastern Railway (LNER) in 1923, Gresley was appointed its Chief Mechanical Engineer, a post he held until his death. He was knighted in 1936.

ENGINEER AND INNOVATOR

Gresley initially started work on the design for a Pacific in 1915, but when his first was actually built in 1922 it was a very different machine. By then Gresley had developed a conjugated valve gear that simplified the drive from three-cylinder engines. Gresley went on to design Britain's largest and most powerful steam locomotive, the 1925 Garratt 2-8-0+0-8-2, and its largest passenger steam locomotive, the Class P2 2-8-2. In an effort to increase efficiency, he also experimented with a high-pressure water-tube boiler originally developed for ships.

From 1928 Gresley developed the A1 Pacifics into A3 Pacifics. These A3s had higher pressure boilers, improving performance further. However, the first recorded steam locomotive speed of 100 mph (161 km/h) was made by an A1 Pacific, *Flying Scotsman*, on 30 November 1934. The next year, Gresley introduced the A4 Pacific, with elegant streamlined styling. It was A4 Pacific *Mallard* that set the current steam locomotive speed record in 1938.

Despite his achievements in steam, Gresley remained open to other methods of rail propulsion and in 1936 began designs for trans-Pennine electrification using 1,500 V DC locomotives. Delayed by World War II, the project was completed in the 1950s.

Record-breaking steam
Mallard was the ultimate evocation of Gresley's A4 Pacific: on 3 July 1938, it set a world steam record speed of 126 mph (203 km/h) that has never been beaten. The locomotive survives at the National Railway Museum, York, England.

The hundred mark
Built in 1923, No. 4472 *Flying Scotsman*
hauled the non-stop London to Edinburgh
service. In 1934 it officially became the
first passenger steam locomotive to
reach a speed of 100 mph (161 km/h).

Streamlined Steam Around Europe

The 1930s was the Golden Age of high-speed, steam-hauled trains in Europe. With national pride at stake, railways competed for the coveted title of the world's fastest train. In Britain the Great Western Railway's *Cheltenham Flyer* was first off the mark in 1932. Hauled by Sir Nigel Gresley's new streamlined A4 Pacifics, the London & North Eastern Railway's *Silver Jubilee* (1935) and *Coronation* (1937) services set new standards in speed, luxury, and reliability. Steam speed records continued to be broken, first by the German Class 05 in 1936 and then by Gresley's *Mallard* in 1938. World War II ended this high-speed excitement, although *Mallard*'s record has never been broken.

△ LNER Class P2, 1934

Wheel arrangement	2-8-2
Cylinders	3
Boiler pressure	220 psi (15.46 kg/sq cm)
Driving wheel diameter	74 in (1,880 mm)
Top speed	approx. 75 mph (121 km/h)

Sir Nigel Gresley's Class P2 locomotives hauled heavy express passenger trains between London and Aberdeen. Six of the powerful engines were built at the London & North Eastern Railway's Doncaster Works between 1934 and 1936. The class was rebuilt as Class A2/2 Pacifics during WWII.

△ LMS Coronation Class, 1938

Wheel arrangement	4-6-2
Cylinders	4
Boiler pressure	250 psi (17.57 kg/sq cm)
Driving wheel diameter	81 in (2,057 mm)
Top speed	approx. 114 mph (183 km/h)

Designed by William Stanier, a total of 38 of these powerful express locomotives were built at the London, Midland & Scottish Railway's Crewe Works between 1937 and 1948. Ten were built with a streamlined casing that was removed after WWII. No. 6229 *Duchess of Hamilton*, refitted with its streamlined casing, has been preserved.

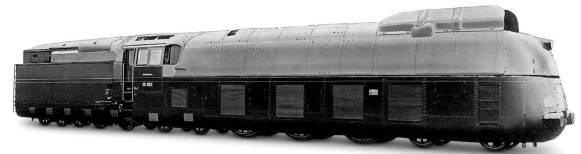

△ DR Class 05, 1935

Wheel arrangement	4-6-4
Cylinders	3
Boiler pressure	290 psi (20.39 kg/sq cm)
Driving wheel diameter	90½ in (2,299 mm)
Top speed	125 mph (201 km/h)

Three of the streamlined Class 05 passenger expresses were built for the Deutsche Reichsbahn in Germany between 1935 and 1937. During 1936 No. 05.002 set a world speed record for steam locomotives of 125 mph (201 km/h) between Berlin and Hamburg. No. 05.001 is preserved in Nürnburg.

▷ LNER Class A4, 1935

Wheel arrangement	4-6-2
Cylinders	3
Boiler pressure	250 psi (17.57 kg/sq cm)
Driving wheel diameter	80 in (2,030 mm)
Top speed	126 mph (203 km/h)

British engineer Sir Nigel Gresley designed the Class A4 streamlined locomotive. Thirty-five of them were built at the London & North Eastern Railway's Doncaster Works between 1935 and 1938. No. 4468 *Mallard* set an unbeaten world speed record for steam engines of 126 mph (203 km/h) on the East Coast Main Line in 1938.

△ SNCB Class 12, 1938

Wheel arrangement	4-4-2
Cylinders	2 (inside)
Boiler pressure	256 psi (18 kg/sq cm)
Driving wheel diameter	82½ in (2,096 mm)
Top speed	103 mph (166 km/h)

The Class 12 was designed by Raoul Notesse for the Belgian state railways. Six of these Atlantic-type locomotives were built between 1938 and 1939 to haul the Brussels to Ostend boat trains. They were retired in 1962 and No. 12.004 has since been preserved.

Travelling Exhibit

The London, Midland & Scottish Railway's streamlined *Coronation Scot* train was shipped across the Atlantic to appear in Baltimore, US. It travelled over 3,000 miles (4,828 km) around the US before being exhibited at the New York World's Fair in 1939. It was unable to return to Britain because of the onset of World War II. The locomotive, No. 6229 *Duchess of Hamilton*, masquerading as No. 6220 *Coronation*, was eventually shipped back to the UK in 1942 but the coaches remained in the US where they were used by the US Army as an officer's mess until after the war, when they too were returned.

Duchess of Hamilton's headlamp One of the two headlamps fitted to *Duchess of Hamilton*, this one remained in the US and is now on display at the Baltimore & Ohio Railroad Museum in Baltimore.

◁ DR Class 03.10, 1939

Wheel arrangement	4-6-2
Cylinders	3
Boiler pressure	290 psi (20.38 kg/sq cm)
Driving wheel diameter	78¾ in (2,000 mm)
Top speed	87 mph (140 km/h)

A total of 60 of these streamlined express passenger locomotives were built for the Deutsche Reichsbahn between 1939 and 1941. After WWII the class was split between East and West Germany and Poland. The German locomotives were rebuilt without their streamline casing, retiring in the late 1970s.

The Silver Jubilee Service

Named to honour the 25-year reign of King George V, the *Silver Jubilee* high-speed express train was introduced by the London & North Eastern Railway between London King's Cross and Newcastle-upon-Tyne in 1935. Painted in two-tone silver and grey, the articulated train was hauled by one of four of Sir Nigel Gresley's new Class A4 streamlined Pacific locomotives named *Silver Link*, *Quicksilver*, *Silver King*, and *Silver Fox*. The service ceased on the onset of World War II.

Inaugural run LNER Class A4 No. 2509 *Silver Link* departs King's Cross station with the inaugural *Silver Jubilee* express to Newcastle on 30 September 1935.

Mallard

During the 1930s the desire to lay claim to the fastest speeds in the world dominated the railways, and records were exchanged between the industrialized nations. Then, on 3 July 1938, one of Sir Nigel Gresley's A4 Class Pacific steam engines, LNER No. 4468 *Mallard*, achieved 126 mph (203 km/h), winning Britain the world record for steam. The start of World War II ended such record attempts, and *Mallard*'s feat has never been beaten.

THE GRESLEY A4 CLASS of streamlined 4-6-2 Pacific locomotives built for the London & North Eastern Railways (LNER) was heralded as an iconic British engine design. The LNER wanted to speed up their services and their Chief Mechanical Engineer, Nigel Gresley, had observed streamlined trains during a trip to Germany. After discussion in 1935, the LNER board gave Gresley and his design team the go-ahead to develop streamliners especially for the railway.

The first of the resulting three-cylinder streamlined A4 Class locomotives, No. 2509 *Silver Link*, was completed at Doncaster Works in September 1935; No. 4468 *Mallard* followed in March 1938. Gresley is said to have modelled their impressive streamlined casing on a wedge-shaped Bugatti railcar he had seen in France. Their futuristic look certainly attracted much publicity. Although the A4 Class was in steam-engineering terms a development of Gresley's earlier A3 Class Pacific, its striking casing could not have made it look more different.

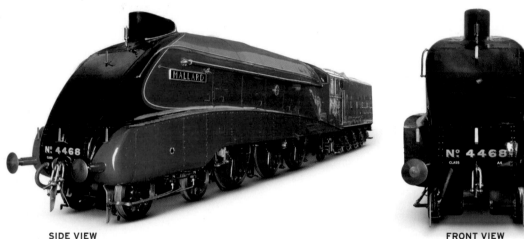

SIDE VIEW

FRONT VIEW

SPECIFICATIONS			
Class	A4	In-service period	1938-63 (*Mallard*)
Wheel arrangement	4-6-2 (Pacific)	Cylinders	3
Origin	UK	Boiler pressure	250 psi (17.57 kg/sq cm)
Designer/builder	Sir Nigel Gresley/Doncaster Works	Driving wheel diameter	80 in (2,030 mm)
Number produced	35 A4s	Top speed	126 mph (203 km/h)

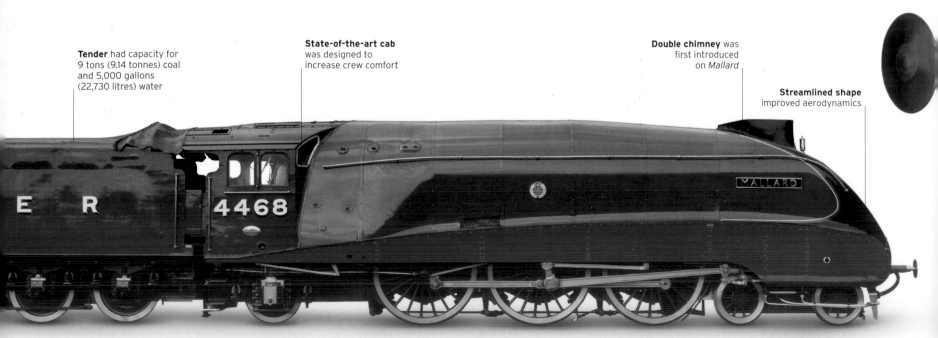

Tender had capacity for 9 tons (9.14 tonnes) coal and 5,000 gallons (22,730 litres) water

State-of-the-art cab was designed to increase crew comfort

Double chimney was first introduced on *Mallard*

Streamlined shape improved aerodynamics

Honouring the speed record
A plaque on the locomotive shows the record speed as 126 mph (203 km/h) – more than 2 miles (3.22 km) per minute. It was measured by timekeepers travelling in a dynamometer car. During the record run, the middle big end ran hot, causing bearing metal to melt.

Serving the Silver Jubilee
The A4 class locomotives were first built to haul the LNER's *Silver Jubilee* high-speed passenger service from London to Newcastle. On the outside, the engine's streamlined shape both improved speed and reduced fuel consumption, while, on the inside, streamlined ports allowed steam to flow freely. *Mallard* was the first A4 to be fitted with the Kylchap exhaust and blast pipe. The sideskirts were removed during WWII to aid maintenance and were replaced only during the later restoration.

MALLARD

Nº 4468

CLASS

EXTERIOR

The streamlined design and smooth casing of *Mallard* not only offered a speed advantage but also helped capture the public imagination. The engine had a wedge-shaped front end with a door built into it to allow access to the smokebox for servicing purposes, in particular for clearing out ash; the door's shape earned it the nickname "cod's mouth". *Mallard's* innovative Kylchap exhaust system was located beneath its double chimney, which was so successful that it led to the whole class of engines being rebuilt with this type of chimney. The A4's unique sideskirts, or valances, were designed by the engineer Oliver Bulleid.

1. Metal nameplate **2.** Engine number and class, hand-painted on the nose **3.** Front buffer **4.** Whistle **5.** Aerodynamic chimney **6.** Coupling hook at front **7.** Brass builder's plaque **8.** Driving wheel **9.** Detail of outside connecting rod big end **10.** Driving wheel return crank **11.** Axle box and cover **12.** Leaf spring suspension

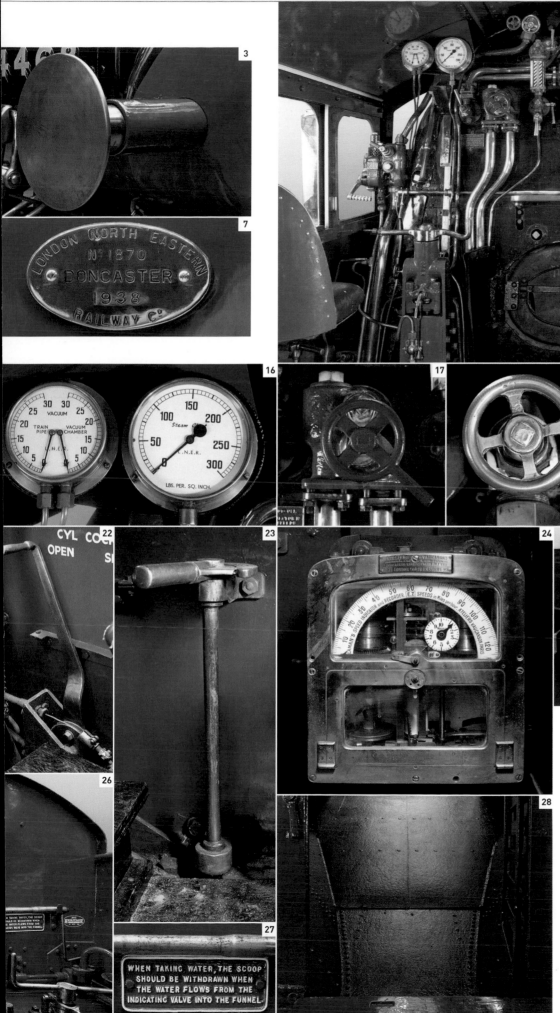

CAB INTERIOR

The locomotive cab was cramped compared to many non-British designs. The crew had to work in concert in the small space to get the most out of the engine. From his bucket seat, the driver controlled the steam delivered to the cylinders using the regulator. The fireman shovelled coal on to the 41¼ sq ft (3.83 sq m) grate area of the firebox, and ensured the boiler contained the right amount of water. Tenders were fitted with a 18-in- (46-cm-) wide corridor so that the engine crews could change over while the train was moving.

13. Cab controls and backhead of boiler **14.** Brake controls
15. Reverser control **16.** Vacuum gauge and steam chest gauge
17. Injector control **18.** Blower shutoff valve, left, pressure gauge
shutoff valve, right **19.** Water-level gauge **20.** Boiler pressure gauge
21. Firebox door **22.** Cylinder cock control **23.** Water control lever
for injectors **24.** Flaman speed recorder **25.** Driver's seat
26. Access door to coal space **27.** Plaque attached to rear of tender
with instructions for use of the water scoop **28.** Tender coal space

The Age of Speed and Style

Symbolized by the futuristic designs of the trains, planes, and automobiles of that period, the decade before World War II could rightly be called "The Age of Speed". Across the world railway companies were introducing modern high-speed expresses designed to entice the travelling public on board with their luxurious interiors, slick service, and dependable fast schedules. Apart from a few diesel-powered streamliners in Germany and the US, these iconic trains were hauled by the latest Art Deco-style steam locomotives, many designed by some of the world's leading industrial designers.

△ VR S Class, 1937

Wheel arrangement 4-6-2

Cylinders 3

Boiler pressure 200 psi (14.06 kg/sq cm)

Driving wheel diameter 73 in (1,854 mm)

Top speed 86 mph (138 km/h)

First introduced in 1928, the four Australian Victoria Railways' S Class Pacific-type locomotives were given a streamlined casing in 1937 to haul the new non-stop, Art Deco-style, *Spirit of Progress* express between Melbourne and Albury. They had all been scrapped by 1954 after the introduction of diesels.

▷ Japan/China Class SL7, 1935

Wheel arrangement 4-6-2

Cylinders 2

Boiler pressure 220 psi (15.5 kg/sq cm)

Driving wheel diameter 78³⁄₄ in (2,000 mm)

Top speed 87 mph (140 km/h)

Built by Kawasaki Heavy Industries in Japan and the Shahekou Plant in the Kwantung Leased Territory in China, the 12 Pashina-type locomotives hauled the *Asia Express* during Japanese control of the South Manchuria Railway between 1934 and 1943. Designated Class Shengli 7 after the war, they remained in service in China until the 1970s.

▽ MILW Class A, 1935

Wheel arrangement 4-4-2

Cylinders 2

Boiler pressure 300 psi (21.09 kg/sq cm)

Driving wheel diameter 84 in (2,134 mm)

Top speed 112¹⁄₂ mph (181 km/h)

Designed to haul the US *Hiawatha* expresses, four of these high-speed Atlantic-type Class A locomotives were built for the Milwaukee Road (MILW) from 1935 to 1937. Locomotive "A" No.2 achieved 112¹⁄₂ mph (181 km/h) between Milwaukee and New Lisbon in May 1935.

△ CN Class U-4-a, 1936

Wheel arrangement	4-8-4
Cylinders	2
Boiler pressure	275 psi (19.33 kg/sq cm)
Driving wheel diameter	77 in (1,956 mm)
Top speed	90 mph (145 km/h)

Five of these streamlined Confederation-type express passenger locomotives were built for Canadian National Railways by the Montreal Locomotive Works in 1936. They remained the premier express locomotives between Toronto and Montreal until replaced by diesels in the 1950s.

TALKING POINT

Rail and Road

By the mid-1930s, American Art Deco-style cars and streamlined steam trains were capable of achieving speeds of 120 mph (193 km/h). Industrial designers such as the American Gordon Buehrig, the Franco-American Raymond Loewy, the Englishman John Gurney Nutting, and Italian-born Frenchman Ettore Bugatti all left their mark on the brief but exciting period of technological progress that ended with the onset of World War II.

Speed rivalry Now highly sought after, Jack Juratovic's iconic "Road and Track" prints of 1935 feature a Duesenberg Torpedo Phaeton car racing a streamlined steam train.

△ NSWGR Class C38, 1943

Wheel arrangement	4-6-2
Cylinders	2
Boiler pressure	245 psi (17.22 kg/sq cm)
Driving wheel diameter	69 in (1,750 mm)
Top speed	80 mph (129 km/h)

Designed in 1939, five of the standard-gauge Australian Class C38 express passenger locomotives were actually delivered to the New South Wales Government Railways by Clyde Engineering of Sydney between 1943 and 1945. After hauling expresses they were retired between 1961 and 1976.

△ PP&L "D" Fireless locomotive, 1939

Wheel arrangement	0-8-0
Cylinders	2
Boiler pressure	130 psi (9.14 kg/sq cm)
Driving wheel diameter	42 in (1,067 mm)
Top speed	20 mph (32 km/h)

Streamlined, but not fast, this Pennsylvania Power & Light Co. fireless shunter was built by Heisler for the Hammermill Paper Co. in Erie. Used in industrial plants where inflammable fuel would be a hazard, fireless locomotives stored steam in their boilers. The largest of this type built, No. 4094-D is on display in the Railroad Museum of Pennsylvania in Strasburg, US.

PIONEERS

Raymond Loewy

Nicknamed "The father of Streamlining", French-born Raymond Loewy (1893–1986) was an American industrial designer known for his wide-ranging work for US industry. In addition to designing world-famous logos for oil companies, such as Shell, and railways, he also left his mark on Studebaker cars and iconic railway locomotives such as the Pennsylvania Railroad's Class K4s, T1 and S1 streamlined steam engines. After opening an office in London in 1930, Loewy went on to restyle the Baldwin Locomotive Co.'s early diesel locomotives. Loewy returned to live in his native France in 1980 and died a few years later.

Standing tall Raymond Loewy stands on one of his iconic designs, Pennsylvania Railroad's unique Class S1 6-4-4-6 experimental streamliner locomotive, the US's largest and fastest high-speed locomotive.

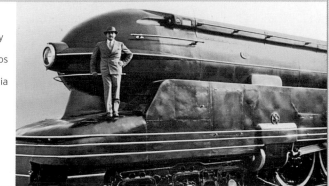

Diesel and Electric Streamliners

The 1930s saw the introduction of high-speed diesel and electric trains in Europe and North America. Designed by leading engineers such as Ettore Bugatti and tested in wind tunnels, these streamliners caught the public's imagination, broke world speed records, and ushered in the new age of high-speed rail travel. In Europe the Germans led the way with their *Flying Hamburger*, the forerunner of today's intercity expresses, and in the US the *Pioneer Zephyr* reached new heights of futuristic modern design. Sadly the onset of World War II brought an abrupt end to this exciting progress.

△ DR Class SVT 137
Fliegender Hamburger, 1935

Wheel arrangement two-car articulated set – front and rear bogies 2′ Bo′ 2′	
Transmission each car electric (1 traction motor)	
Engine each car Maybach 12-cylinder diesel 8,850 cc	
Total power output 810 hp (604 kW)	
Top speed 99 mph (160 km/h)	

With a prototype built in 1932, the Deutsche Reichsbahn train entered service in 1935 between Berlin and Hamburg; it had a buffet and seated 98. The diesel-electric *Fliegender* (flying) *Hamburger* established the world's fastest regular train service with an average speed of 77 mph (124 km/h). Inactive during WWII, it saw service in France in 1945–49, then returned to operate in Germany until 1983.

△ SBB Class Ae8/14, 1931

Wheel arrangement (1′A)A1A(A1′) + (1A′)A1A(A1′)	
Power supply 15 kV 17 Hz AC, catenary	
Power rating 7,394–10,956 hp (5,514–8,173 kW)	
Top speed 62 mph (100 km/h)	

Three prototype Class Ae8/14 electric locomotives were built for the Swiss Federal Railways' (Schweizerische Bundesbahnen, or SBB) Gotthard line in the 1930s. Each of these powerful double locomotives had eight driving axles and could haul heavy trains unaided over this difficult route. No. 11852 (shown) was for a time the most powerful locomotive in the world.

▷ Bugatti railcar (autorail), 1932/33

Wheel arrangement each car 2 x 8-wheel bogies, 2 or 4 axles powered	
Transmission mechanical	
Engine each car 2 or 4 Bugatti 12,700 cc	
Total power output 4 engines 800 hp (596 kW)	
Top speed 122 mph (196 km/h)	

Designed by Ettore Bugatti and built in the Bugatti factory in Alsace, France, these petrol-engined railcars were supplied as single-, double-, or triple-car units. The most comfortable and fastest was the 48-seat, two-car, four-engined "Presidentiel", which set a world rail-speed record of 122 mph (196 km/h) in 1934.

▷ GWR streamlined railcar, 1934

Wheel arrangement 2 x 4-wheel bogies, 1 powered	
Transmission mechanical	
Engine 8,850 cc AEC diesel	
Total power output 130 hp (97 kW)	
Top speed approx. 63 mph (100 km/h)	

First introduced by the Great Western Railway in 1934, these streamlined diesel railcars were nicknamed "Flying Bananas" and remained in service on British Railways until the early 1960s. Production versions, including parcels cars and articulated buffet sets, were fitted with two AEC diesel engines allowing a top speed of 80 mph (129 km/h).

TECHNOLOGY

German Experiment

The *Schienenzeppelin*, or "rail zeppelin", was an experimental railcar with an aluminium body, which looked like a Zeppelin airship. The front-end design of this prototype bore an uncanny resemblance to the Japanese *Bullet Train* of the 1960s.

Weighing only 20 tons (20.32 tonnes), this 85-ft (26-m) long propeller-driven car was powered by a BMW 12-cylinder petrol aircraft engine producing a power of 600 hp (447 kW). In June 1931 it set a world land-speed record for rail vehicles using air propulsion when it reached 143 mph (230 km/h) on the Berlin to Hamburg line. The railcar was scrapped in 1939 to provide material for the German war effort in World War II.

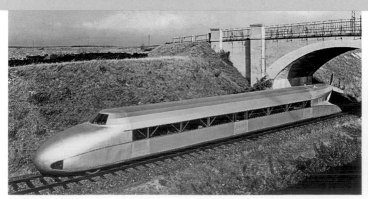

Zeppelin train Built by Franz Kruckenberg of Hannover the *Schienenzeppelin* only had two axles and was designed to carry 40 passengers.

Rear fairing The wind-tunnel–designed fairing had a four-bladed propeller made of ash wood.

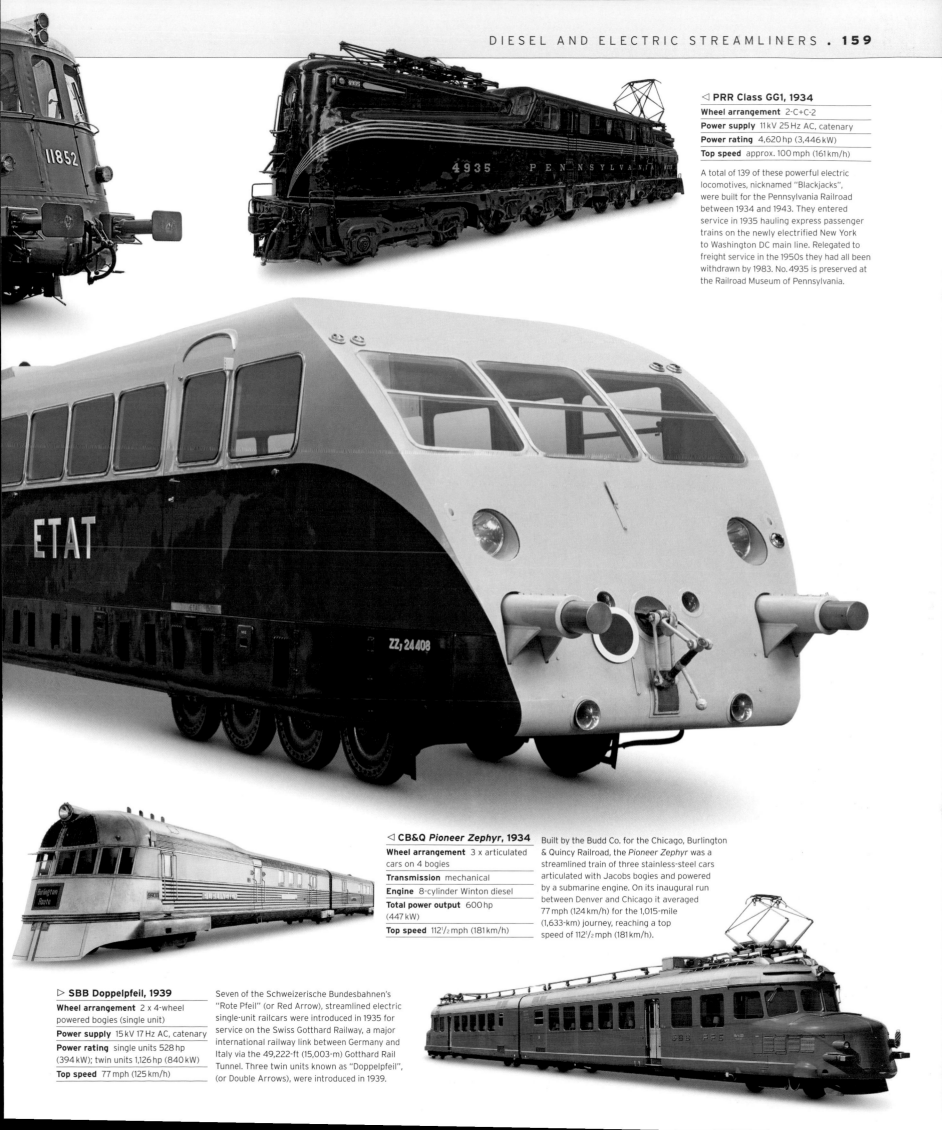

◁ **PRR Class GG1, 1934**

Wheel arrangement	2-C+C-2
Power supply	11 kV 25 Hz AC, catenary
Power rating	4,620 hp (3,446 kW)
Top speed	approx. 100 mph (161 km/h)

A total of 139 of these powerful electric locomotives, nicknamed "Blackjacks", were built for the Pennsylvania Railroad between 1934 and 1943. They entered service in 1935 hauling express passenger trains on the newly electrified New York to Washington DC main line. Relegated to freight service in the 1950s they had all been withdrawn by 1983. No. 4935 is preserved at the Railroad Museum of Pennsylvania.

◁ **CB&Q Pioneer Zephyr, 1934**

Wheel arrangement	3 x articulated cars on 4 bogies
Transmission	mechanical
Engine	8-cylinder Winton diesel
Total power output	600 hp (447 kW)
Top speed	112¹⁄₂ mph (181 km/h)

Built by the Budd Co. for the Chicago, Burlington & Quincy Railroad, the *Pioneer Zephyr* was a streamlined train of three stainless-steel cars articulated with Jacobs bogies and powered by a submarine engine. On its inaugural run between Denver and Chicago it averaged 77 mph (124 km/h) for the 1,015-mile (1,633-km) journey, reaching a top speed of 112¹⁄₂ mph (181 km/h).

▷ **SBB Doppelpfeil, 1939**

Wheel arrangement	2 x 4-wheel powered bogies (single unit)
Power supply	15 kV 17 Hz AC, catenary
Power rating	single units 528 hp (394 kW); twin units 1,126 hp (840 kW)
Top speed	77 mph (125 km/h)

Seven of the Schweizerische Bundesbahnen's "Rote Pfeil" (or Red Arrow), streamlined electric single-unit railcars were introduced in 1935 for service on the Swiss Gotthard Railway, a major international railway link between Germany and Italy via the 49,222-ft (15,003-m) Gotthard Rail Tunnel. Three twin units known as "Doppelpfeil", (or Double Arrows), were introduced in 1939.

Practical Diesels and Electrics

World War I had left Europe's railways in tatters; coal was scant and expensive, and, while steam was still popular, other forms of traction would soon emerge to herald the end of an era. In mountainous countries such as Italy and Switzerland, an abundance of clean and cheap hydroelectric power made possible the electrification of main lines. Powerful electric locomotives, such as the Swiss "Krocodils", were soon hauling heavy trains over demanding routes, while in Italy speed records were being broken on Mussolini's new high-speed railway. At the other end of the scale, small diesel and electric shunters (or switchers) were being introduced in Europe and North America as a more efficient way of marshalling trains.

△ GIPR Class WCP 1, 1930

Wheel arrangement	1'Co2'
Power supply	1.5 kV DC, catenary
Power rating	2,158 hp (1,610 kW)
Top speed	75 mph (121 km/h)

The first electric locomotives to be used in India, 22 of these powerful passenger engines were built from 1930 by Metropolitan-Vickers in the UK for the Great Indian Peninsula Railway. The first of these, No. 4006 *Sir Roger Lumley*, is on display at the National Rail Museum, New Delhi.

▷ SBB Class Ce 6/8 II and Ce 6/8 III, 1919-20

Wheel arrangement	1-C+C-1
Power supply	1.5 kV AC, catenary
Power rating	3,647 hp (2,721 kW)
Top speed	47 mph (76 km/h)

Serving until 1980, 51 of these electric engines were built to haul heavy freight on the Swiss Federal Railways' (Schweizerische Bundesbahnen, or SBB) Gotthard line from 1919 to 1927. Their long noses, for which they were nicknamed "Krokodils" (crocodiles), contain the motors.

▷ DR E04, 1933

Wheel arrangement	1'Co1'
Power supply	15 kV AC 17 Hz, catenary
Power rating	2,694 hp (2,010 kW)
Top speed	75 mph (121 km/h)

A total of 23 Class E04 electric locomotives were built for Deutsche Reichsbahn for service on the newly electrified Stuttgart to Munich main line. Members of the class stayed in service in West Germany until 1976 and in East Germany until 1982. Several of these have been preserved.

▷ GHE T1, 1933

Wheel arrangement	A1 (0-2-2)
Transmission	mechanical
Engine	4-cylinder diesel
Total power output	123 hp (92 kW)
Top speed	25 mph (40 km/h)

This unique four-wheel 3-ft 3-in- (1-m-) gauge diesel railcar (or *Triebwagen*) was built by Waggonfabrik Dessau in 1933 for the Gernrode-Harzgeroder Railway in Germany. After WWII it became No. 187.001 of the East German Deutsche Reichsbahn and was used as a workman's tool wagon. Seating 34 passengers, this restored railcar runs on the Harz narrow-gauge railways

△ PRR Class B1, 1934

Wheel arrangement	C (0-6-0)
Power supply	11 kV AC, catenary
Power rating	697 hp (520 kW)
Top speed	25 mph (40 km/h)

Fourteen of these single-unit electric switchers were built at Altoona Works by the Pennsylvania Railroad in 1934. They spent most of their life performing empty carriage movements in and out of Penn Station in New York, US, before retiring in the early 1970s.

△ DR Class Kö, 1934

Wheel arrangement	B (0-4-0)
Transmission	mechanical
Engine	79 hp (959 kW) diesel as modified
Total power output	18-22 kW (24-29 hp)
Top speed	11 mph (18 km/h)

These small diesel mechanical shunters, known as *Einheitskleinlokomotiven*, served at small stations on the Deutsche Reichsbahn. Fitted with only a foot brake, some were converted to run on LPG during WWII. Three of these, including No. 199.011 shown, have been converted to operate as Class Kö II on the 3-ft 3-in- (1-m-) gauge Harz railways.

▷ LMS Diesel Shunter No. 1831, 1931

Wheel arrangement	C (0-6-0)
Transmission	hydraulic
Engine	Davey Paxman 6-cylinder diesel
Total power output	400 hp (298 kW)
Top speed	25 mph (40 km/h)

This was the first experimental diesel-hydraulic shunter in the UK. It was built by the London, Midland & Scottish Railway at its Derby Works in 1931 using the frame and running gear of a Midland Railway 1377 Class 0-6-0 steam locomotive of the same number. It was not successful and was officially withdrawn in 1939.

◁ FS Class ETR 200, 1937

Wheel arrangement	3-car articulated on 4-wheel bogies
Power supply	3 kV DC, catenary
Power rating	1,408 hp (1,050 kW)
Top speed	126 mph (203 km/h)

Entering service between Milan and Naples in 1937, a total of 18 of these three-car electric multiple units were built by Breda for the Italian state railways. The streamlined shape was designed after wind tunnel tests, and in July 1939 unit ETR 212 set a world record for electric rail traction of 126 mph (203 km/h). The class was in regular service until the 1990s, and ETR 212 has since been preserved.

TECHNOLOGY

Track Inspection

During the 19th century the maintenance of thousands of miles of railway track, often in places inaccessible by road, was only made possible by teams of gangers walking the lines or travelling on unpowered handcars (also known as pump cars or jiggers). These were propelled by pushing a wooden arm up and down. By the 20th century more ingenious methods had been introduced, such as motorized road vehicles fitted with flanged wheels. Road-rail inspection vehicles are still used today on remote railways around the world. In the US these are known as hi-rail vehicles; in Scotland, Land Rovers are adapted for use on the West Highland Line.

Buick Ma&Pa Car No. 101, 1937 Originally used as a funeral car, this vehicle was converted to run on the Maryland & Pennsylvania Railroad to test a radio communication system between locomotives and the railway's offices.

Reading MU No. 800

The Reading Company (also known as the Reading Railroad) was a railroad and coal mining conglomerate that expanded rapidly from the 1830s. The company developed commuter rail services from Philadelphia, building the imposing Reading Terminal station in 1893. The decision to electrify many of the commuter lines was taken in 1928 and, despite the Wall Street Crash of 1929, the expansion continued and electric trains began running from July 1931.

THE READING MULTIPLE UNIT (MU) cars were specially built for the electrification project. Incorporating the latest technology, the cars were designed to be cheap to run. Aluminium was used extensively in the car body to make them light and reduce the amount of electricity needed to operate them. The MU cars were designed to work on their own or as part of much longer trains, as they had cabs at both ends. Furthermore, they were simpler to operate and much quicker than the steam locomotives they had replaced.

The first 61 cars were ordered in 1928 and delivered in 1931. The company ordered more cars as the electric network continued to expand; by 1933 more than 84 miles (135 km) of the system was electrified. Some cars remained in service for 60 years, including 38 that were rebuilt between 1963 and 1965 and survived until 1990. Most of the older cars were withdrawn a year or two after the state-government-owned Southeastern Pennsylvania Transportation Authority (SEPTA) took control of services on the former Reading Company lines in 1983.

FRONT VIEW

REAR VIEW

Electric and steam
The Reading Railroad used the "Reading Lines" brand name for its passenger services. In addition to running electric trains, the railway operated several steam locomotives.

SPECIFICATIONS			
Class	EPa/EPb	**In-service period**	1931-90 (No. 800)
Wheel arrangement	B2	**Railway**	Reading Railroad
Origin	USA	**Power rating**	480 hp (358 kW)
Designer/builder	Harlan & Hollingsworth	**Power supply**	11 kV AC 25 HZ overhead lines
Number produced	91 Reading MU cars	**Top speed**	70 mph (113 km/h)

Pantograph draws electrical power from overhead wires

Ventilator mounted on roof

Windows fitted to full length of passenger car

Electric bus connector transfers power to next car

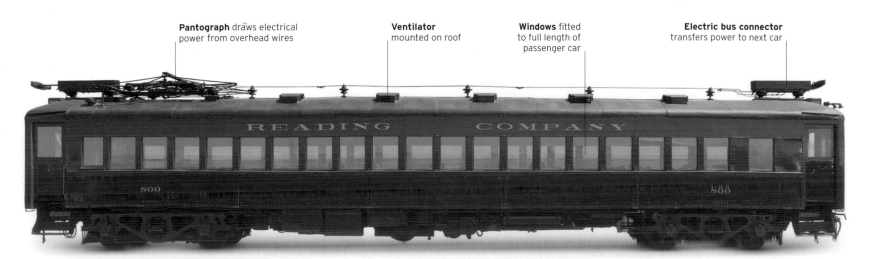

Electric bus connector
Visible above the headlight, the electric bus connector was first used in the US on the Reading Co. MU cars. The connectors on adjacent coaches touched, enabling electricity to pass safely from one car to the next. As a result, only a single pantograph was required to operate several motor cars at once.

EXTERIOR

Looking similar to today's commuter rail cars, the MU car – with its steel body sides, driving cabs at either end, and automatic doors – was technologically advanced when introduced. High-voltage power lines ran along the vehicle roof from the pantograph (which made contact with the overhead wire power supply) to the electric bus connectors at each end of the car.

1. Railcar number on side **2.** Electric bus connector at each end of car **3.** Headlight **4.** Horn **5.** Insulator for 11 kV AC electric power cable running along roof **6.** Pantograph in lowered position **7.** Marker light **8.** Cab window and wiper blade **9.** Cow catcher **10.** Sockets for 12-pin multiple working cables **11.** 12-pin multiple working cables **12.** Brake shoe **13.** Open journal box showing bearing **14.** Leaf spring suspension on wheel unit **15.** *Taylor Flexible Truck* bogie **16.** Handbrake chain **17.** Air brake control valves

CAB INTERIOR

The driver of a Reading Company MU car had to stand up in the cab, using simple controls developed from those used on electric tramways. They also had to estimate the train's speed, as the early trains were not fitted with speedometers. However, the MU was fitted from new with a cab signalling system that delivered electrical signals to the train via the track.

18. Driver's cab **19.** Handbrake **20.** Ratchet for handbrake **21.** Throttle control, with socket for operation by Allen key **22.** Light switch boxes (left) and brake pressure gauges **23.** Marker light lens (red) indicates end of train **24.** Marker light lens (yellow, shows as white when lit by oil lamp) indicates unscheduled train **25.** Marker light lens (blue, shows as green when lit by oil lamp) indicates scheduled train **26.** Top of brake unit **27.** Cab interior door lock

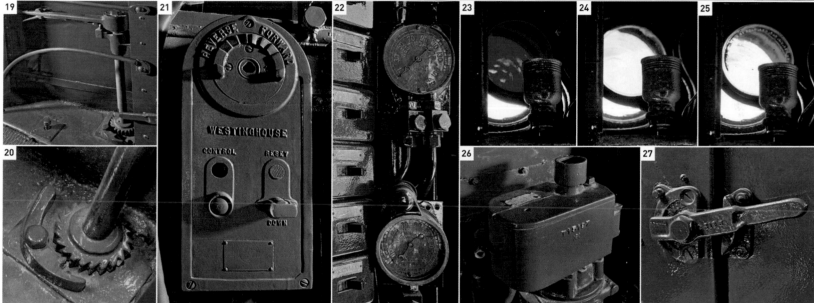

CAR INTERIOR

The standard MU car had 18 rows of 2+2 reversible seats with space for gangway-facing seats and a conductor's office. The freight cars had a separate compartment at one end and fewer seats. The railway also introduced innovations such as internal doors that automatically closed when the train was in motion and thermostats for carriage heating, which reduced costs and improved passenger comfort.

28. Interior of railcar carriage **29.** Light fitting in ceiling **30.** Metal luggage rack above seats **31.** Brass window sash clip **32.** Foot rest and heater under seat **33.** Door to toilet **34.** Emergency axe in glass unit

1940-1959
WAR AND PEACE

WAR TRAFFIC MUST
COME FIRST

DON'T WASTE TRANSPORTATION

ASSOCIATION OF AMERICAN RAILROADS IN COOPERATION WITH THE OFFICE OF DEFENSE TRANSPORTATION

WAR AND PEACE

During World War II railways formed an integral part of the struggle for victory for both Allied and Axis forces. However, rail traffic not directly related to the war effort was discouraged in this period. Many of Europe's railways were devastated during the conflict, and were also much maligned by association for their transport of millions of people to concentration camps. But emerging from the shadows of war, Western Europe's railways returned with a new glamorous face, the Trans-Europ Express (TEE) – part of a major international effort to rebuild and rebrand railways in the war's aftermath.

Although the US had rolled out steam giants such as the "Big Boy" in the early 1940s, the shift to diesel had already started with a rapid transition to the new form of traction. During the post-war period throughout the world, diesel- and electric-power increasingly replaced steam, which was seen as dirty, labour-intensive, and outmoded. The supremacy of the new technologies was enhanced in 1955 when two French electric trains broke the world speed record.

In the UK, the newly nationalized British Railways stuck with steam power until after the publication of the Modernisation Plan in the mid-1950s. Nevertheless, the ever-increasing use of diesel-shunting engines showed the way forward, and in 1955 the prototype *Deltic* appeared, presenting the new face of express transport. However, by the late 1950s, neither the US nor Europe was developing the most revolutionary form of rail transport. That honour went to Japan.

△ **Red carpet train**
A 1941 poster advertises the famous *20th Century Limited* US passenger train, which later featured in the film *North By Northwest*.

> " The railroads ... can be **reached at any moment** by military orders. **Nothing**, therefore, **can replace the railroads**"
>
> ERNST MARQUARDT, GERMAN MINISTRY OF TRANSPORTATION, 1939

◁ **Propaganda poster (1939-45) by Fred Chance,** who worked as an illustrator in Philadelphia and New York

Key Events

▷ **1941** US's Union Pacific Railroad launches its giant "Big Boy" steam locomotives.

▷ **1942** Germany's Class 52 "Kriegslok" (war loco) is introduced as a stripped-down war-time design. Its reliability also helps the post-war reconstruction.

▷ **1945** Allied forces use "train-busters" to destroy German locomotives.

△ **Casualty of war**
This German locomotive was found upended by Allied forces when they captured the town of Muenster, Germany on 11 April 1945.

▷ **1948** Railways are nationalized in the UK. Private companies are replaced by British Railways.

▷ **1949** With the formation of West Germany, the country's railways become the Deutsche Bundesbahn. East Germany's railways keep the name Deutsche Reichsbahn.

▷ **1951** British Railways launches a new range of "Standard" steam designs.

▷ **1954** 1 December, British Railways Modernisation Plan announces the elimination of steam.

▷ **1955** French electrics BB 9004 and CC 7107 reach 206 mph (331 km/h) – a new world record.

▷ **1957** The pan-European Trans-Europ Express network is launched – and a series of iconic trains are built to run on its routes.

World War II Logistics

The transportation of raw materials, troops, military equipment, and ammunition by rail was of strategic importance to the warring powers in World War II. As a result, cheaply constructed powerful freight locomotives – mass-produced in Germany, Britain, and the US – saw active service in war zones. After the war many ran on European national railways – as replacements or as war reparations. A large number of engines, built for the United States Army Transportation Corps (USATC), were sent to Asia under lease-lend agreements and, after the war, by the UN Relief & Rehabilitation Administration.

◁ **LMS 8F, 1935**

Wheel arrangement	2-8-0
Cylinders	2
Boiler pressure	225 psi (15.82 kg/sq cm)
Driving wheel diameter	56 $\frac{1}{4}$ in (1,430 mm)
Top speed	approx. 50 mph (80 km/h)

Designed by William Stanier for the London, Midland & Scottish Railway, these were the standard British freight locomotives for part of WWII. They saw service for Britain's War Department in Egypt, Palestine, Iran, and Italy – 25 were sold to Turkey in 1941. Of the 852 built, some remained in British service until 1968, while Turkish examples ran into the 1980s.

△ **DR Class 52 "Kriegslok", 1942**

Wheel arrangement	2-10-0
Cylinders	2
Boiler pressure	232 psi (16.3 kg/sq cm)
Driving wheel diameter	55 in (1,400 mm)
Top Speed	50 mph (80 km/h)

Around 7,000 of these Deutsche Reichsbahn heavy freight locomotives were built mainly for service on the Eastern Front. A small number remain in service in Bosnia even today, while many, like this Class 52 No. 52.8184-5 rebuild, have been preserved.

◁ **USATC S160, 1942**

Wheel arrangement	2-8-0
Cylinders	2
Boiler pressure	225 psi (15.82 kg/sq cm)
Driving wheel diameter	56$\frac{3}{4}$ in (1,440 mm)
Top speed	approx. 45 mph (72 km/h)

Of the 2,120 austerity Consolidation-type heavy freight locomotives built for the USATC, 800 were shipped to Britain for use in Europe after D-Day. After the war they saw service on many European railways as well as in North Africa, China, India, and North and South Korea.

▷ **USATC S100, 1942**

Wheel arrangement	0-6-0T
Cylinders	2
Boiler pressure	210 psi (14.8 kg/sq cm)
Driving wheel diameter	54 in (1,370 mm)
Top speed	approx. 35 mph (56 km/h)

Built for the USATC, 382 of these locomotives were shipped to Britain and used in Europe after the D-day landings of June 1944. Britain's Southern Railway later bought 15 as shunters.

△ **Class V36 Shunter, 1937**

Wheel arrangement	0-6-0
Transmission	hydraulic
Engine	Deutsche Werke/MAK diesel
Total power output	360 hp (268 kW)
Top speed	approx. 37 mph (60 km/h)

Fitted with four axles but only three pairs of driving wheels, these diesel locomotives were built for the German armed forces (Wehrmacht) and were used for shunting duties. They saw widespread use in Europe and North Africa after the war.

▷ **Armoured Car, 1942**

Type	4-wheel
Capacity	130 (whole train)
Constuction	armour-plated steel
Railway	German Wehrmacht

This camouflaged car formed part of a German Wehrmacht BP42 armoured train that protected supply and transport trains in the Balkans and Russia. An armoured Class 57 0-10-0 steam locomotive was positioned in the centre of the train, which consisted of a combination of infantry, navigating, anti-aircraft, and artillery wagons, with converted tank turrets.

▷ SR Class Q1, 1942

Wheel arrangement	0-6-0
Cylinders	2 (inside)
Boiler pressure	230 psi (16.17 kg/sq cm)
Driving wheel diameter	61 in (1,550 mm)
Top Speed	50 mph (80 km/h)

Designed by Oliver Bulleid for the Southern Railway, these freight locomotives were lightweight, which enabled them to operate over most of the company's network. A total of 40 were built, and they all remained in service on the Southern Region of British Railways until the 1960s. This is No. C1, the first of the series.

The Maryland Car

In 1947, US journalist Drew Pearson set out to help the people of war-stricken France and Italy. A Friendship Train travelled around the US gathering $40 million of relief supplies. In response, the French sent a Merci (thank you) Train filled with gifts back to the US. Arriving in New York in 1949, the train consisted of a series of European boxcars used to transport soldiers and horses during the war. There were 49 cars – one for each US state (although the District of Columbia and Hawaii had to share). The Maryland Car, shown here, was originally built for the Paris, Lyon & Mediterranean Railway in 1915. It is now on display at the Baltimore & Ohio Railroad Museum, Baltimore.

△ WD Austerity, 1943

Wheel arrangement	2-8-0
Cylinders	2
Boiler pressure	225 psi (15.82 kg/sq cm)
Driving wheel diameter	56¼ in (1,430 mm)
Top speed	approx. 45 mph (72 km/h)

Designed by R.A. Riddles for the British War Department, these freight trains were "austerity", or cheaper, versions of the LMS 8F. Of the 935 built, many saw service in mainland Europe after D-day in June 1944. After the war, 733 were in operation for British Railways, while others worked in the Netherlands, Hong Kong, and Sweden.

▽ Indian Class AWE, 1943

Wheel arrangement	2-8-2
Cylinders	2
Boiler pressure	210 psi (14.76 kg/sq cm)
Driving wheel diameter	61½ in (1,562 mm)
Top speed	approx. 62 mph (100 km/h)

These huge locomotives were built by Baldwin Locomotive Works for the USATC for hauling heavy freight trains in India during WWII. They were fitted with 7-ft- (2,134-mm-) diameter boilers and 40 became Indian Railways Class AWE. One of these, No. 22907 *Virat*, has been restored to working order at the Rewari Steam Loco Shed.

DR No. 52.8184-5

Built to serve Germany during World War II, the Deutsche Reichsbahn (DR) Class 52 "Kriegslok" had a simple design and was constructed from materials that were easy to source during wartime. Nevertheless, it became a rugged classic vital to many countries long after the conflict ended, partly because it could haul heavy loads on lightweight track. Although designed to last only a few years, the class also proved very durable.

THE CLASS 52 "KRIEGSLOK" (*Kriegs-Dampflokomotive*, or war steam locomotive) came out of Germany's need to construct locomotives quickly during World War II, while maintaining maximum production capacity for armaments. The plan was to build 15,000 locomotives, with production spread throughout occupied Europe. Only around 7,000 were actually made, but Germany's Class 52 remains one of the most numerous classes ever built. The locomotive shown on these pages was built in Vienna in 1944.

The ten driving wheels gave the 52 enough grip to pull 1,968 tons (2,000 tonnes) across a level surface at 31 mph (50 km/h). In addition, the locomotive included plenty of cold weather protection, useful in a war that progressed into Russia in winter. Some were even equipped with tenders that could recycle exhaust steam back into water, meaning they could travel long distances without having to refill.

After the war "Kriegsloks" remained in service. Some were modernized and renumbered, including the engine now known as No. 52.8184-5, which is kept in Stassfurt, Germany.

FRONT VIEW

REAR VIEW

Deutsche Reichsbahn

Manufacturing for war
Like much of German industry during World War II, the Deutsche Reichsbahn was harnessed to the war effort. The Class 52 "Kriegslok" epitomized the machines of war built during that period.

SPECIFICATIONS			
Class	52 or Kriegs-Dampflokomotive 1	In-service period	1942–present (No. 52.8184-5)
Wheel arrangement	2-10-0	Cylinders	2
Origin	Germany	Boiler pressure	232 psi (16.3 kg/sq cm)
Designer/builder	Hauptausschuß Schienenfahrzeuge	Driving wheel diameter	55 in (1,400 mm)
Number produced	approx. 7,000 Class 52	Top speed	50 mph (80 km/h)

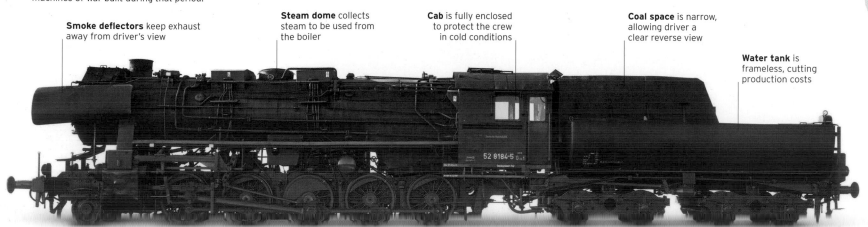

Smoke deflectors keep exhaust away from driver's view

Steam dome collects steam to be used from the boiler

Cab is fully enclosed to protect the crew in cold conditions

Coal space is narrow, allowing driver a clear reverse view

Water tank is frameless, cutting production costs

Simple elegance
Paring the design to the bare
minimum helped give the
Class 52 its austere look.
Essential components were
simplified wherever possible.

52 8184-5

H. Unt
03.04
V

Wartime Service

The importance of the railways in World War II can be measured by the efforts made on both sides to destroy networks. Between 1940 and 1942, Germany's air force, the Luftwaffe, launched more than 10,000 attacks on Britain's rail network, but failed to prevent it from transporting the fuel, food, equipment, and munitions that the nation needed. Germany's railways not only transported vital goods, but also played a part in history's worst act of genocide – the transportation of Jews and other groups to death camps.

On the Allied side, the Trans-Iranian Railway in the Middle East kept oil flowing from the Persian Gulf to the Soviet Union, while in North America trains delivered supplies to Atlantic ports for shipping to Britain.

THE BIG BUILD

By 1942, with the tide turning in the Allies' favour, planning began for the invasion of Axis-occupied Europe. Recognizing that railways would be crucial to success, Britain and the US began a programme of locomotive building on an unprecedented scale. By D-Day, 6 June 1944, more than a thousand new engines had been built to haul supply trains, and they were immediately put into service. The Allies achieved victory in May 1945 and, while this was due to many factors, there is no doubt that railways played a critical part.

The Women's Voluntary Service (WVS) filled the roles left vacant by the 110,000 British railwaymen who served in the war. Here, WVS members clean the engines at a London Midland & Scottish Railway depot.

US Moves into Diesel

The diesel locomotive represented the greatest advance in US railways during the 20th century. Although diesel engines had been around since the 1890s, the challenge was to make them small and light enough to fit within the confines of a locomotive, yet powerful enough to haul a train. The breakthrough had come in 1935 when General Motors unveiled their 12-cylinder, 2-cycle engine that was 23 per cent smaller and, thanks to lightweight alloys, 20 per cent lighter than its predecessors. Accelerating into the 1940s the diesel began to conquer the US.

◁ **Boxley Whitcomb 30-DM-31, 1941**

Wheel arrangement	0-4-0
Transmission	mechanical
Engine	8-cylinder Cummins
Total power output	150 hp (120 kW)
Top speed	approx. 20 mph (32 km/h)

Built by the Whitcomb company of Rochelle, Illinois, "30" referred to the locomotive weight range in tons and "DM" to its transmission (diesel-mechanical). The Boxley Materials Co. of Roanoke, Virginia bought No. 31 from the Houston Shipbuilding Corporation of Texas in 1953.

◁ **VC Porter No. 3, 1944**

Wheel arrangement	A1-A1
Transmission	electric
Engine	not known
Total power output	300 hp (224 kW)
Top speed	approx. 20 mph (32 km/h)

H.K. Porter Inc. was one of the largest manufacturers of industrial locomotives in the US – by 1950, it had built 8,000. This rod-driven switcher Porter No. 3, built for the Virginia Central Railroad, is the last of the 28 of its type built. It is now preserved at the Virginia Museum of Transportation.

▷ **PMR GM EMD SW-1 No. 11, 1942**

Wheel arrangement Bo-Bo

Transmission electric

Engine EMD Model 567 V-6 engine

Total power output 600 hp (448 kW)

Top speed 45 mph (72 km/h)

The Pere Marquette Railway, which took its name from a 17th-century French Jesuit priest, served the Great Lakes region of the US and Canada. The EMD SW-1 Class was introduced in 1936, but No. 11 was delivered to the railway in April 1942 to begin shunting hoppers at Eireau, Ontario. It was retired in 1984.

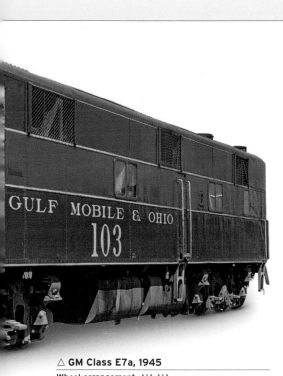

△ **GM Class E7a, 1945**

Wheel arrangement A1A-A1A

Transmission electric

Engine 2 x EMD Model 567A, 12-cylinder

Total power output 2,000 hp (1491 kW)

Top speed 85 mph (137 km/h)

Supplied to over 20 railways, General Motors's E7 was one of the first standard American diesels. Between 1945 and 1949, 428 of the E7a cab units were produced along with 82 E7b boosters. The Gulf, Mobile, & Ohio Railroad's No. 103, shown here, was notable for appearing in the film *The Heat of the Night*.

△ **Baldwin Class DS-4-4-660, 1946**

Wheel arrangement Bo-Bo

Transmission electric

Engine 4-cycle engine

Total power output 660 hp (492 kW)

Top speed 60 mph (96 km/h)

The Chesapeake & Western (C&W, known locally as the "Crooked and Weedy") operated over 53$^1/_2$ miles (86 km) of the Shenandoah Valley (US). In 1946 it took delivery of three diesel units. With running costs at 25 cents per mile (as opposed to the 96 cents for steam), it marked a turning point for the CHW. No. 662 was retired in 1964 and languished in a scrapyard before being donated to the Virginia Museum of Transportation.

◁ **Ma&Pa GM EMD Type NW2, 1946**

Wheel arrangement Bo-Bo

Transmission electric

Engine 12-cylinder engine

Total power output 1,000 hp (750 kW)

Top speed 60 mph (97 km/h)

This type was first introduced in 1939. The Maryland & Pennsylvania Railroad (a short line linking Baltimore with York and Hanover, Pennsylvania) took delivery of Nos. 80 and 81, two Type NW2 switchers built by the Electro-Motive Division of General Motors, in December 1946. In total, 1,145 NW2s were shipped between 1939 and 1949 to over 50 railways (in contrast to the Ma&Pa's modest pair, Union Pacific bought 95). No. 81 has been part of the Railroad Museum of Pennsylvania's collection since 1997.

Post-war US

Some railways needed persuading that diesel could match the haulage power of steam, but, once General Motors's freight demonstrator and prototypes had convinced them, the economic argument was irresistible. Mainline locomotives fell into two categories: cab units and hood units. The former, with their sleek bodywork and colourful liveries, handled the expresses and were augmented by boosters for extra power. In the hood unit, the workhorse, the engine (or engines), radiators, and ancillary equipment were mounted on a platform above the chassis with the cab placed at one end or in the centre. The transition from steam to diesel was accomplished within 20 years; by 1960 around 34,000 diesel locomotives operated in the US.

▷ B&A GE 70-ton switcher, 1946

Wheel arrangement	Bo-Bo
Transmission	electric
Engine	2 x Cooper-Bessemer FDL-6T 6-cylinder 4-cycle engines
Total power output	660 hp (492 kW)
Top speed	60 mph (96 km/h)

The Baltimore & Annapolis Railroad was mainly a commuter line that in 1950 succumbed to road competition and replaced passenger trains with buses. That year it bought a solitary diesel, a General Electric 70-ton switcher – No. 50 – for freight operations. The type was introduced in 1946 as a lighter, low-cost option for secondary routes, and 238 were built up to 1955. Retired in 1986, No. 50 is preserved at the Baltimore & Ohio Museum.

△ Baldwin S12 switcher, 1950

Wheel arrangement	Bo-Bo
Transmission	electric
Engine	De Lavergne Model 606A SC 4-cycle engine
Total power output	1,200 hp (895 kW)
Top speed	60 mph (96 km/h)

Employing a turbocharged version of the powerful Model 606A engine, the S12 switcher was famous for its hauling prowess, as demonstrated by Baldwin's original No. 1200. Here, masquerading as No. 1200, is Earle No. 7 or, in the records of its operators, the United States Navy, No. 65-000369. The USN took 18 of the 451 S12s shipped between 1951 and 1956 and stationed this unit at its ordnance depot in Earle, New Jersey.

▷ B&O F7 Class, 1949

Wheel arrangement	Bo-Bo
Transmission	electric
Engine	EMD 567B 16-cylinder engine
Total power output	1,500 hp (1,119 kW)
Top speed	50-120 mph (80-193 km/h)

The F7 was the most numerous of the General Motors's F Series; 2,341 A units and 1,467 B (booster) units were built by 1953. The speed variation was a product of eight different gear ratios. Though tailored for freight, many US railways used F7s for front-line passenger services until the 1970s. No. 7100, shown, was bought by the Baltimore & Ohio Railroad in 1951 and enjoyed a second career on the Maryland Area Regional Commuter (MARC) system from 1987 to the late 1990s.

◁ N&W EMD GP9 Class, 1955

Wheel arrangement	Bo-Bo
Transmission	electric
Engine	EMD 567C 16-cylinder engine
Total power output	1,750 hp (1,305 kW)
Top speed	75 mph (125 km/h)

General Motors's "Geep Nine" remains one of the most successful and long-lasting of diesels, although not the most attractive. Looks did not count for US and Canadian railways, which between them bought 4,087 A units and 165 type B boosters from 1954 to 1963. No. 521 was one of 306 GP9s on the books of the Norfolk & Western, and many remain in service on secondary lines and with industrial users; some Class 1 railways still use them as shunters.

△ **Budd RDC railcar, 1949**

Wheel arrangement	Bo-2
Transmission	mechanical
Engine	2 x General Motors Type 6-110 6-cylinder engines
Total power output	275 hp (205 kW)
Top speed	85 mph (137 km/h)

After WWII the Budd Co. used its expertise in building lightweight stainless-steel carriages to assemble diesel railcars (or multiple units) for secondary and local passenger services. A prototype Rail Diesel Car (RDC) was unveiled in 1949 and impressed with its economy. By 1962, 398 were in operation. Out west, RDCs provided a stopping service over the 924 miles (1,487 km) between Salt Lake City, Utah, and Oakland, California. The RDC was also exported to Australia, Brazil, Canada, and France.

◁ **N&W ALCO T6 (DL440) Class, 1958**

Wheel arrangement	Bo-Bo
Transmission	electric
Engine	ALCO 251B 6-cylinder 4-cycle engine
Total power output	1,000 hp (746 kW)
Top speed	60 mph (96 km/h)

The American Locomotive Co. (ALCO) introduced the T6 (the "T" stood for "Transfer") in 1958 believing there was a demand for a switcher capable of shuttling trains between yards and terminals at higher speeds. This was not the case and up to 1969 just 57 had been delivered, of which the Norfolk & Western Railway took 38. Retired in 1985, No. 41 is kept at the Virginia Museum of Transportation, US.

N&W GP9 Class No. 521

The last major US rail operator to switch from steam to diesel was the Norfolk & Western Railway (N&W), based in Roanoke, Virginia. As part of its drive to eliminate steam, the N&W already ran electric trains on some of its routes. From 1955 it moved to diesel, first with ALCO RS-3s, and then bought 306 model GP9 diesel-electric locomotives from General Motors's Electro-Motive Division (EMD). Most of the GP9s were destined for freight duties but some, including No. 521, hauled passenger trains.

THE EMD GENERAL PURPOSE (GP) road switcher diesel engines first appeared in 1949 and became the most successful range of mid-power diesels in North America. The first model, the GP7, was built from 1949 to 1954, when the improved GP9 version was introduced. The "Geeps", as the locomotives were nicknamed, were bought in large numbers to replace the steam locomotives still in use during the 1950s. Continuously updated, the last "Geep" model was the GP60 produced until 1994.

Locomotives 501 to 521 were the last GP9s bought by the N&W and were equipped with steam boilers to heat passenger coaches. At first they replaced the fast J Class steam locomotives that worked the N&W passenger trains in the 1950s, but when the passenger services ceased, they were used for freight alongside the 285 other GP9s operated by this railroad. In 1982, the N&W merged with the Southern Railway to become Norfolk Southern Railway, which is today one of the largest Class 1 railways in the US.

FRONT VIEW

REAR VIEW

Special logo
For the introduction of the new passenger GP9 diesels the N&W logo in yellow was unusually mounted on a round plate with a black background on the locomotive front.

SPECIFICATIONS			
Class	GP9	**In-service period**	1958–85 (No. 521)
Wheel arrangement	BoBo	**Transmission**	electric
Origin	USA	**Engine**	EMD 567C 16-cylinder
Designer/builder	General Motors EMD	**Power output**	1,750 hp (1,305 kW)
Number produced	306 (GP9s for N&W)	**Top speed**	75 mph (125 km/h)

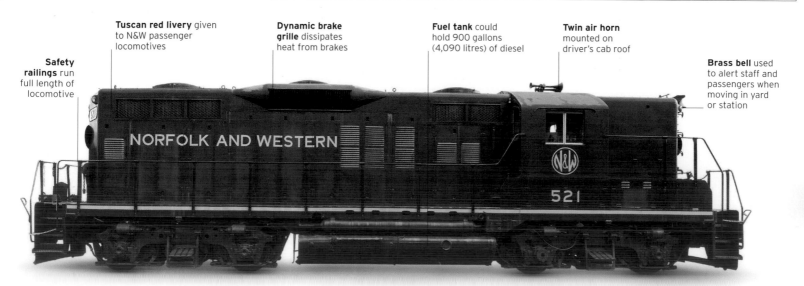

Tuscan red livery given to N&W passenger locomotives

Dynamic brake grille dissipates heat from brakes

Fuel tank could hold 900 gallons (4,090 litres) of diesel

Twin air horn mounted on driver's cab roof

Safety railings run full length of locomotive

Brass bell used to alert staff and passengers when moving in yard or station

"The redbirds"
The last 21 GP9s bought by the N&W for
passenger trains were given a special livery
of Tuscan Red with yellow lettering, earning
them the nickname "the redbirds".

EXTERIOR

The GP9 was a simple but rugged design with features in common with all EMD locomotives of the time, such as the standard US "knuckle" coupler, originally developed in the 1890s and fitted at each end of the GP9. The use of spare parts that were interchangeable between EMD models was one of the reasons so many of these locomotives were sold in the 1950s.

1. Numberplate on front end of engine **2.** Twin headlights **3.** Ladder to access top of engine **4.** Electrical connection cap **5.** "Knuckle" coupler **6.** Diesel fuel cap **7.** Emergency fuel cut-off **8.** Front steps **9.** Wheel unit (bogie) **10.** Air horn positioned above cab **11.** Spring on engine bogie **12.** Air-brake cylinder **13.** Dynamic brake grille **14.** Clasp brake **15.** Brass bell on front end **16.** Door to cab

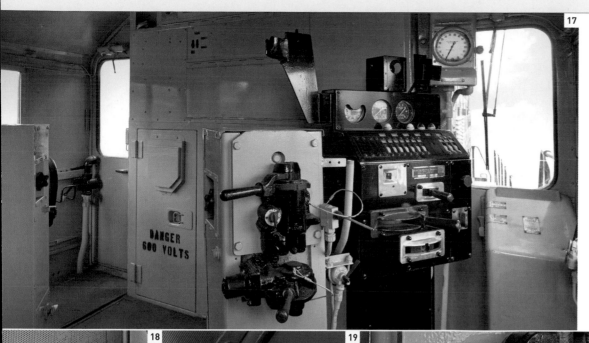

CAB INTERIOR

The driver's cab had a standard EMD control station, with lever-operated power and reverse and braking controls. The locomotive and train brake equipment were located alongside each other. The locomotive could be driven in either direction at full speed, and the driver had a good view forward from each end of the cab. The power controller (or throttle) had eight "notches", so the driver could increase or decrease power gradually.

17. Interior of cab with engineer's controls **18.** Emergency brake valve **19.** Windshield wiper motor **20.** Switches for windshield wipers **21.** Brake control levers **22.** Warning sign **23.** Speedometer **24.** Control panel circuit breaker switches **25.** Air brake gauges **26.** Load indicator **27.** Power controller

Britain Makes the Change

By the 1940s the rail system in Britain consisted of four major companies and many smaller light railways. In 1948 the "Big Four" and the majority of the smaller railways were nationalized under one umbrella company – British Railways. The new company commissioned a report to look at ways of stemming the losses they were incurring as a result of competition from air and road traffic. Known as the Modernisation Plan and published on 1 December 1954, the report made a number of recommendations, including the replacement of all steam engines. Tests in the late 1950s with "pilot-scheme" diesels were intended to demonstrate which locomotives to order in quantity. Orders for thousands of new diesels would follow in the next decade.

△ BR (W) Gas Turbine No. 18000, 1949

Wheel arrangement	A1A-A1A
Transmission	electric
Engine	Brown Boveri Gas Turbine
Total power output	2,500 hp (1,865 kW)
Top speed	90 mph (145 km/h)

This revolutionary locomotive was delivered to British Railways in 1949 from Switzerland and was used for 10 years on the BR Western Region. In 1965 it left the UK and was used for research in Switzerland and Austria, returning in 1994 to the UK where it is now preserved.

△ BR Class 08, 1953

Wheel arrangement	0-6-0
Transmission	electric
Engine	English Electric 6KT
Total power output	350 hp (261 kW)
Top speed	20 mph (32 km/h)

Based on a wartime design of diesel shunter ordered by the London, Midland & Scottish Railway, over 950 Class 08 locomotives were built by five British Railways workshops between 1953 and 1959. Smaller batches of similar locomotives using different engines were also built. Sixty years on some remain in service. No. 08 604 *Phantom* is preserved at Didcot, UK.

△ BR Class 05, 1954

Wheel arrangement	0-6-0
Transmission	mechanical
Engine	Gardner 8L3
Total power output	201 hp (150 kW)
Top speed	17 mph (27 km/h)

This engine was one of several designs of smaller shunting locomotives delivered to British Railways in the 1950s. Later classified as Class 05, 69 were built between 1954 and 1961. Few remained in service for more than a decade as the freight traffic they were built for disappeared after the BR network was reduced following the Beeching Report.

◁ English Electric prototype *Deltic*, 1955

Wheel arrangement	Co-Co
Transmission	electric
Engine	2 x Napier Deltic D18-25 engines
Total power output	3,300 hp (2,460 kW)
Top speed	106 mph (171 km/h)

Built speculatively by English Electric, *Deltic* was the prototype for the 22 Type 5 *Deltic* D9000 Class 55 diesel locomotives bought for services on the East Coast route from London to York and Edinburgh. They were to replace the famous London & North Eastern Railway design A4 Pacific steam engines.

▷ BR Type 1 Class 20, 1957

Wheel arrangement Bo-Bo
Transmission electric
Engine English Electric 8SVT MkII
Total power output 986 hp (735 kW)
Top speed 75 mph (121 km/h)

This class was one of the most successful of all the Modernisation Plan locomotives. A total of 227 were built for British Railways between 1957 and 1968. The class saw limited passenger services but could work in multiples and, coupled together, could handle heavy traffic. Some remain in use with UK freight operators nearly 60 years later.

◁ BR Class 42, 1958

Wheel arrangement B-B
Transmission hydraulic
Engine 2 x Maybach MD650 engines
Total power output 2,100 hp (1,566 kW)
Top speed 90 mph (145 km/h)

These locomotives were based upon successful V200 engines that ran in West Germany and used the same engines as their German cousins. Known as *Warships*, they were used by British Railways principally on the Western Region from London Paddington to Devon, Cornwall, and South Wales until withdrawn from service in 1972. This is No. 801 *Vanguard*.

△ BR Class 108, 1958

Wheel arrangement 2-coach multiple unit
Transmission mechanical
Engine 2 x BUT/Leyland 6 cylinder
Total power output 300 hp (224 kW)
Top speed 70 mph (113 km/h)

British Railways's Modernisation Plan led to the replacement of steam locomotives, and more than 4,000 diesel multiple units were ordered. These new self-propelled "Derby Lightweight" trains were much cheaper to operate than the steam trains they replaced.

▷ BR Type 4 Class 40, 1958

Wheel arrangement 1Co-Co1
Transmission electric
Engine English Electric 16SVT MkII
Total power output 1,972 hp (1,471 kW)
Top speed 90 mph (145 km/h))

This class was designed to replace the fastest steam locomotives working express trains initially between London and Norwich and later all over the UK. The initial pilot batch of 10 was expanded to a final class of 200 by 1962.

Deltic Prototype

During its time, the English Electric prototype *Deltic*, first tested in 1955, was the most powerful diesel locomotive in the world. Using Napier Deltic engines developed to power fast naval patrol boats, *Deltic* produced high levels of performance while weighing less than most contemporary locomotives. British Railways ordered 22 production *Deltics* in 1958, introducing 100 mph (161 km/h) express trains to the UK.

THE NAPIER DELTIC ENGINE had a unique layout with three banks of six cylinders in a triangular formation. To enable each group of cylinders to work efficiently, the crankshaft for one group operated in the opposite direction to the other two – the resulting opposed piston engine was both compact and very powerful. Its light weight meant that two derated naval engines could be installed in a six-axle locomotive; the power available made the *Deltic* the most powerful diesel locomotive of its time.

The *Deltic* prototype began tests with British Railways in 1955, initially on the West Coast route from London to Liverpool and Carlisle. The locomotive remained the property of its builders – English Electric – whose engineers accompanied it on every trip. From 1959 it operated on the East Coast route from London to York and Edinburgh. It was here that *Deltic* would excel, leading to an order for 22 locomotives, with a slightly smaller bodyshell, which would replace 55 express passenger steam engines. Retired from use in 1961, the prototype was presented to London's Science Museum and today is part of the UK's National Railway Museum collection.

SIDE VIEW

FRONT VIEW

Major manufacturer
The Deltic prototype was built and owned by English Electric. Founded in 1918 this major British engineering company built hundreds of diesel and electric engines until 1968.

SPECIFICATIONS	
Class	Deltic prototype
Wheel arrangement	Co-Co
Origin	UK
Designer/builder	English Electric, Vulcan Foundry
Number produced	1
In-service period	1955–61
Transmission	electric
Engine	2 x Napier Deltic D18-25
Power output	3,300 hp (2,460 kW)
Top speed	106 mph (171 km/h)

A spacious cab was provided at either end of the locomotive

Engine room housing the two Napier Deltic engines and two generators

Nameplate is unusual, as the prototype had no number, just the name

Bright blue, cream, and gold livery is unique to the prototype and has not been seen on a British locomotive before or since

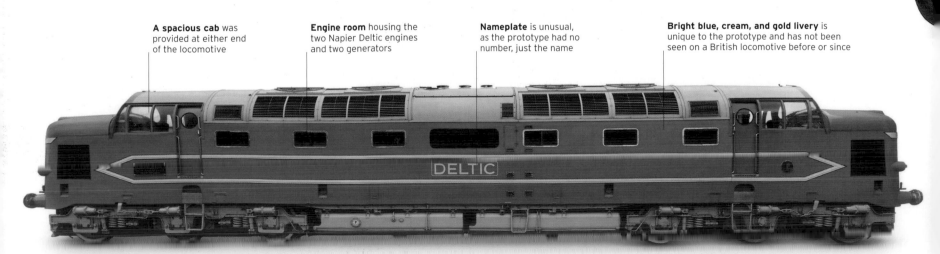

DELTIC

American appeal
The headlight was part of the rounded North American styling for the prototype locomotive. In practice, this headlight was never used because the locomotive never left the UK for trials anywhere else.

THE ENGLISH **EE** ELECTRIC Co LTD

EXTERIOR

The *Deltic* prototype was built with export markets around the world in mind and consequently used styling similar to the US streamlined diesel designs that had been in use since the 1940s. The bright blue, cream, and gold livery made it stand out from every other locomotive in the UK when it started trials in 1955. The design had many innovative features for its era – such as retractable steps, streamlined lights, and buffers.

1. Painted name plaque 2. Large headlight space (light never fitted) 3. Streamlined electric marker light 4. Front buffer 5. Front coupling hook 6. Horn bracket 7. Windscreen and wiper blades 8. Sandbox 9. Folding chrome steps 10. Air brake chain 11. Exhaust vent positioned at centre of engine 12. Metal steps up to driver's door 13. Leaf spring suspension 14. Fuel gauge 15. Shed shore supply (electricity)

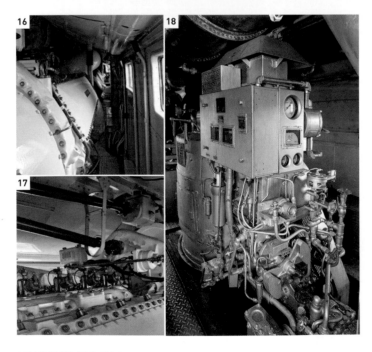

ENGINE ROOM

The design for the compact Deltic engine originated in World War II Junkers aeroplane engines from Germany. The engine is made from aluminium alloy and designed to be as lightweight as possible. To fit the engines into the locomotive was an engineering challenge, as the loading gauge (maximum height and width) of UK trains is smaller than on European railway systems.

16. Inside Deltic Engine 17. Controls at top of engine 18. Steam heating boiler

CAB INTERIOR

The locomotive had identical cabs at either end, each designed to give a good view forward, from a raised position, through a two-piece windscreen. This clear view of the line ahead was essential for safe operation at a speed of 100 mph (161 km/h). The locomotive was operated by a two-man crew, one driving and the other monitoring ancillary equipment, such as steam heating.

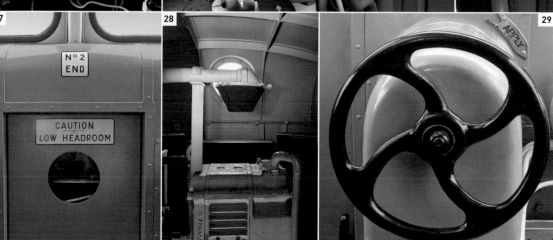

19. Left side of cab   **20.** Right side of cab   **21.** Warning light attached to ceiling   **22.** Westinghouse vacuum brake   **23.** Wiper motor   **24.** Driver's display panel   **25.** Loco brake (above) and power handle (below)   **26.** Orange electricity conduits   **27.** Maintenance doorway to the nose   **28.** Vacuum exhauster in the nose   **29.** Wheel brake   **30.** Steam heating control

Europe Follows the US

As Europe emerged from the chaos and damage inflicted by World War II, many railway companies based their future planning on the US where diesels had been replacing steam for nearly a decade. A wide variety of manufacturers using an equally wide choice of diesel engines built locomotives for state railways across Europe. Labour-intensive steam was replaced with diesels, which were cheaper to run although more expensive to buy. The process was gradual in most countries; some steam engines survived until 1977 in West Germany, and they never entirely disappeared in East Germany.

▽ DB V200 (Class 220), 1954

Wheel arrangement	B-B
Transmission	hydraulic
Engine	2 x Maybach MD 650 engines
Total power output	2,170 hp (1,618 kW)
Top speed	87 mph (140 km/h)

Designed to replace steam locomotives on heavy express passenger trains in the mid-1950s, the Class 220s were displaced to less important routes by electrification in the 1960s and 70s. All were withdrawn by the Deutsche Bundesbahn by 1984, but many went on to work for other operators in Greece, Switzerland, and Italy.

△ NSB Class Di3, 1955

Wheel arrangement	Co-Co
Transmission	electric
Engine	EMD 16-567-C
Total power output	1,750 hp (1,305 kW)
Top speed	65 mph (105 km/h)

The Swedish firm Nydqvist & Holm AB (NoHAB) built diesel locomotives under license for the major US diesel locomotive builder EMD, then owned by General Motors. As well as the Di3 locomotives, delivered in two types to Norwegian State Railways (Norges Statsbaner AS, or NSB), similar locomotives were supplied to Denmark and Hungary. The locomotives remain in service with freight operators in several European countries.

Diesel Shunters

While the big mainline diesel engines attracted attention, using diesel locomotives in shunting yards was just as transformational. While labour-intensive steam machines needed a team of operatives and had to be kept "in steam" even when at rest, diesel shunters could be operated by one person, and simply switched off when not in use. Crew conditions were better too and, in many cases, so was the visibility from the cab. The advantages of the diesels were recognized even before World War II, and after the conflict their use became more and more widespread. Many of the 1950s designs had long working lives.

△ SNCF Class C61000, 1950

Wheel arrangement	0-6-0
Transmission	electric
Engine	Sulzer 6 LDA 22
Total power output	382 hp (285 kW)
Top speed	37 mph (60 km/h)

Ordered immediately after WWII in 1945, but not delivered until 1950–1953, the 48 C61000 locomotives were used for shunting in freight yards and for short-distance freight. Twelve of the locomotives were used with coupled powered "slave" units to double the power available for shunting.

△ DB VT11.5 (Class 601/602), 1957

Wheel arrangement B'2+2'2'+2'2'+2'2'+
2'2'+2'2'+2'B

Transmission hydraulic

Engine 2 x MTU engines

Total power output 2,060 hp (1,536 kW)

Top speed 100 mph (160 km/h)

The Class 601s were First Class only diesel-powered train sets used for Trans-Europ Express services from 1957 to 1972, reaching Paris, Milan, Amsterdam, and Ostende. Some were rebuilt as Class 602 from 1970 with 2,1450 hp (1,600 kW) gas turbines in place of the two diesel engines. The train sets were withdrawn from service in 1990.

△ SNCF Class CC6500, 1957

Wheel arrangement Bo-Bo

Transmission electric

Engine 2 x SACM MGO VSHR V12

Total power output 1,824 hp
(1,360 kW)

Top speed 81 mph (130 km/h)

Because of their shape these locomotives were nicknamed "Sous-marin" (submarines). Twenty were delivered to SNCF to replace steam locomotives in the west of France where they worked until the 1980s; all were withdrawn by 1988. Tested widely when new, their builder Alsthom also exported the design – 37 to Algeria and 25 to Argentina.

◁ DB VT98 (Class 798), 1955

Wheel arrangement single-car
rail bus

Transmission mechanical

Engine 2 x Büssing AG U10 engines

Total power output 295 hp (220 kW)

Top speed 56 mph (90 km/h)

These rail bus vehicles were introduced in West Germany from 1953 to 1962 – initially the single-engined VT95 version, and then this more powerful two-engined VT98 version. In total 913 powered and 1,217 unpowered trailer cars (of both types) replaced steam locomotives on many rural lines across West Germany.

◁ PKP Class SM30, 1957

Wheel arrangement Bo-Bo

Transmission electric

Engine Wola V-300

Total power output 295 hp (220 kW)

Top speed 37 mph (60 km/h)

This was the first diesel-electric locomotive designed and built in Poland – its initial models used an engine originally designed for army tanks. Ultimately 909 of the locomotives were built by Fablok in Chrzanów in southern Poland between 1956 and 1970, many for industrial users. Polish State Railways (or PKP) received 302. Some are still in use in 2014.

△ DR V15 (Class 101), 1959

Wheel arrangement 0-4-0

Transmission hydraulic

Engine 6 KVD 18 SRW

Total power output 148 hp (110 kW)

Top speed 22 mph (35 km/h)

The East German V15 (and later V18) diesel shunters were built in large numbers for both the Deutsche Reichsbahn and industrial rail operators such as mines and steelworks. Built in Potsdam by VEB Lokomotivbau Karl Marx Babelsberg, many were also exported to other Eastern Bloc countries.

Great Journeys
The Blue Train

The Blue Train is one of the world's most luxurious trains. Styling itself as a "hotel-on-wheels", the train travels 994 miles (1,600 km) between Pretoria and Cape Town in South Africa and passes through scenery that ranges from lush vineyards to rugged semi-desert.

THE PREDECESSORS OF *The Blue Train* came into service in the 1890s, picking up passengers from the Union-Castle liners docking in Cape Town and transporting them to the gold and diamond fields in the north. These early trains soon began catering to prospectors and wealthy travellers by offering more comfortable rail experiences. By 1923 the luxury Cape Town to Johannesburg trains were called the Union trains. The *Union Express* travelled from Cape Town to Johannesburg while the *Union Limited* made the return journey. By 1928 these trains offered facilities such as hot and cold water and heated carriages, later acquiring dining saloons in 1933 and air-conditioning in 1939. The trains' distinctive blue livery was introduced in 1936.

World War II caused train services to be suspended in 1942. They resumed in 1946, the same year that "those blue trains", as they were popularly known, adopted *The Blue Train* as their official designation. The trains have since been completely rebuilt twice, once in the 1970s and once in the 1990s.

Today the soundproofed, carpeted compartments all feature their own en-suite bathrooms (luxury suites include full-sized bathtubs). The train has underfloor heating, a restaurant car offering fine dining, two lounge cars, an observation car (which converts to a conference car), as well as a 24-hour butler service and a laundry service.

Departing Cape Town
The Blue Train leaves the Cape to head north to Pretoria, flanked by the famous profile of Table Mountain.

THE BLUE TRAIN INSIGNIA

The decor of each coach is unique, with birchwood panelling, marble finishes, and gold-plated fittings throughout.

THE ROUTE TODAY

The Blue Train at one time travelled all the way to Victoria Falls in Zimbabwe, but this route has since been discontinued. Several others are now available but only as chartered services. The train's standard route from Cape Town to Pretoria runs through the Cape winelands and under the spectacular Hex River Mountains, where the train emerges from a series of tunnels into the arid region known as the Klein Karoo (Little Karoo). Here the train makes a stop at Matjiesfontein, a town that sprang up in 1884 around a refreshment station for passing trains, and which remains preserved in its Victorian state. The service then continues on to Pretoria, passing through the semi-desert landscape of the Great Karoo. On the return journey from Pretoria, passengers may disembark to visit the mining town of Kimberley, site of the diamond rush that began in the 1870s, before journeying on to Cape Town.

Lounging in luxury
There are two lavishly appointed lounges aboard the train, where passengers can expect five-star service. Once on board, food and

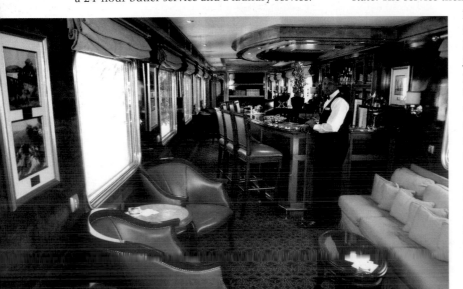

KEY FACTS

DATES
1946 *The Blue Train* name is formally adopted
1970s, 1997 The train is refurbished

TRAINS
Train Set 1 Charter train; 14 carriages accommodate 52 passengers
Train Set 2 Cape Town-Pretoria-Cape Town train; 19 carriages accommodate 80 passengers
Locomotives 2 x 14E Class electric locomotives, dual current; 118 tons (120 tonnes)
Carriages 9 ft 5 in (2.9 m) wide - 2 in (50 mm) wider than standard South African rolling stock. Thinner steel sides allow greater interior space
Speed 49 mph (80 km/h) with a maximum of 86 mph (138 km/h)
Weight Complete train 98 ½ tons (100 tonnes)

JOURNEY
Cape Town to Pretoria (weekly) 994 miles (1600 km) 27 hours
Cape Town to Durban (biannually, Sept & Nov) 473 miles (760 km) 21 hours. Also available for charter

RAILWAY
Gauge Cape Gauge 3 ft 6 in (1,067 mm)
Tunnels Four Hex River Tunnels: twin tunnel 1,640 ft (500 m); single tunnels 3,609 ft (1,100 m), 3,937 ft (1,200 m) and 44,291 ft (13.5 km)
Bridges Orange River Station Bridge; Vaal River Crossing (Warrenton)
Highest point 5,751 ft (1,753 m), Johannesburg

Matjiesfontein 2
This quaint and tiny museum town, little more than a single street, is now primarily a tourist destination.

The Hex River Valley 1
The train passes through the vineyards of the valley and through four tunnels beneath the Hex (Witch) River Mountains, named for the girl who haunts them in local legend.

Great Karoo Desert

Table Mountain

Matjiesfontein

Paarl
Worcester
George
Cape Town

Kaaimans River 5
The route crosses the Kaaimans estuary and passes through seven tunnels.

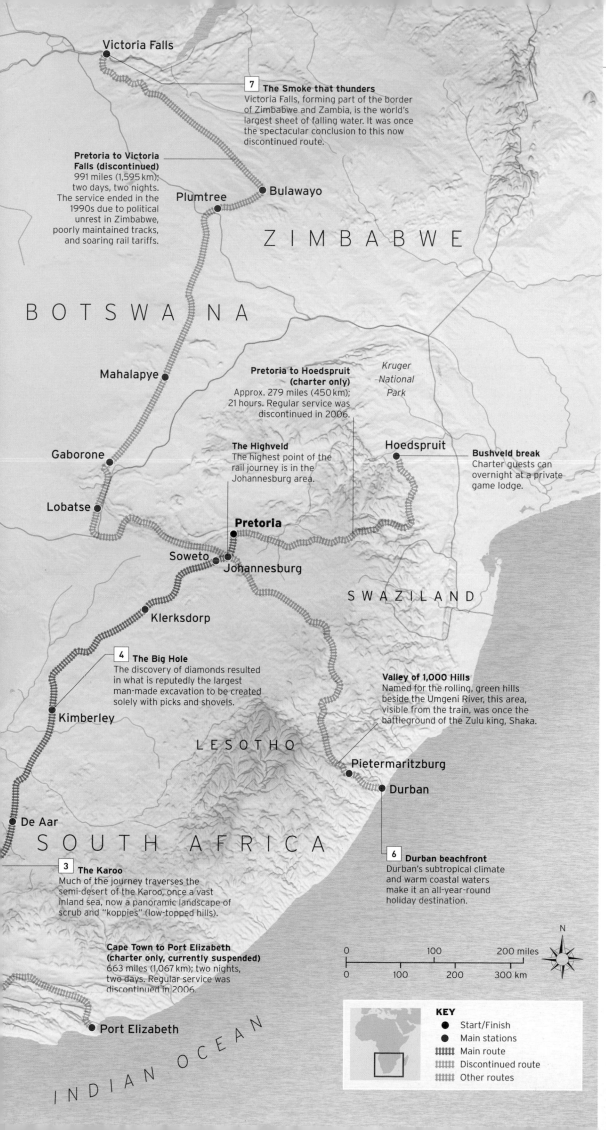

Victoria Falls

7 **The Smoke that thunders**
Victoria Falls, forming part of the border of Zimbabwe and Zambia, is the world's largest sheet of falling water. It was once the spectacular conclusion to this now discontinued route.

Pretoria to Victoria Falls (discontinued)
991 miles (1,595 km); two days, two nights. The service ended in the 1990s due to political unrest in Zimbabwe, poorly maintained tracks, and soaring rail tariffs.

Plumtree • **Bulawayo**

Z I M B A B W E

B O T S W A N A

Mahalapye

Kruger National Park

Pretoria to Hoedspruit (charter only)
Approx. 279 miles (450 km); 21 hours. Regular service was discontinued in 2006.

The Highveld
The highest point of the rail journey is in the Johannesburg area.

Hoedspruit

Bushveld break
Charter guests can overnight at a private game lodge.

Gaborone

Lobatse

Pretoria

Soweto
Johannesburg

S W A Z I L A N D

Klerksdorp

4 **The Big Hole**
The discovery of diamonds resulted in what is reputedly the largest man-made excavation to be created solely with picks and shovels.

Valley of 1,000 Hills
Named for the rolling, green hills beside the Umgeni River, this area, visible from the train, was once the battleground of the Zulu king, Shaka.

Kimberley

L E S O T H O

Pietermaritzburg

Durban

De Aar

S O U T H   A F R I C A

6 **Durban beachfront**
Durban's subtropical climate and warm coastal waters make it an all-year-round holiday destination.

3 **The Karoo**
Much of the journey traverses the semi-desert of the Karoo, once a vast inland sea, now a panoramic landscape of scrub and "koppies" (low-topped hills).

Cape Town to Port Elizabeth (charter only, currently suspended)
663 miles (1,067 km); two nights, two days. Regular service was discontinued in 2006.

Port Elizabeth

I N D I A N   O C E A N

| 0 | | 100 | | 200 miles |
| 0 | 100 | 200 | | 300 km |

KEY
● Start/Finish
● Main stations
▥ Main route
▥ Discontinued route
▥ Other routes

AN AFRICAN JOURNEY

The Blue Train travels through a range of terrain, from the lush Cape to the arid Karoo. The stopping points on the journey reflect South Africa's colonial past and the source of the country's wealth at the height of its powers.

Electric Charge

In the early part of the 20th century several European railways had already started to use electric rather than steam locomotives on main lines in the Swiss and Austrian Alps – they were among the first to use this powerful new technology. Plans to expand electrified railways were delayed almost everywhere in Europe by World War II, which led to the destruction of much railway infrastructure. As post-war rebuilding got underway, most European countries turned to electrified railways, and the 1950s saw new electric trains being widely introduced.

△ **BR Class 70 No.20003, 1948**

Wheel arrangement Co-Co

Power supply 750 V DC third rail, overhead lines

Power rating 2,200 hp (1,641 kW)

Top speed 75 mph (120 km/h)

Following two similar locomotives (CC1/CC2) delivered to Southern Railway in 1941, No. 20003 was built at the Ashford Locomotive Works, Kent, in 1948 for British Railways. Like the earlier two it was used until the late 1960s, mainly on the London to Brighton main line and other Sussex routes.

◁ **BLS Ae 4/4, 1944**

Wheel arrangement Bo-Bo

Power supply 15 kV AC, 16²/₃ Hz, overhead lines

Power rating 3,950 hp (2,946 kW)

Top speed 78 mph (126 km/h)

Designed and built in Switzerland during WWII, the Ae 4/4 design was revolutionary, using a light steel body mounted on two-axle bogies. It produced nearly 4,000 hp (2,984 kW), which was the equivalent of two or three steam engines. These design principles have been used for electric locomotives ever since.

▽ **SNCF Class BB9000, 1954**

Wheel arrangement Bo-Bo

Power supply 1,500 V DC, overhead lines

Power rating 4,000 hp (2,983 kW)

Top speed 206 mph (331 km/h)

This was one of two pairs of experimental express passenger engines using two-axle bogies that were delivered to the French state railways in 1952-54: BB9003 and 9004 were built in France by Jeumont Schneider. On 29 March 1955 BB9004, along with CC7107, set a world record of 206 mph (331 km/h) for locomotives, which was not beaten until 2006.

▷ BR Class EM1/ Class 76, 1954

Wheel arrangement Bo-Bo

Power supply 1,500 V DC, overhead lines

Power rating 1,868 hp (1,393 kW)

Top speed 65 mph (105 km/h)

Built for the electrification of the Manchester to Sheffield route via Woodhead, the first prototype was made for British Railways in 1940 but remained unused owing to WWII. It was tested in the Netherlands from 1947 to 1952, and was returned when the Woodhead line's electrification was completed.

◁ FS Class ETR, 1952

Wheel arrangement 7-car EMU

Power supply 3,000 kV DC, overhead lines

Power rating 3,487 hp (2,600 kW)

Top speed 124 mph (200 km/h)

Featuring a driving cab on the roof, and a panoramic lounge at the front with just 11 First Class seats, the "Settebello" (Seven of Diamonds – named after an Italian card game) was the epitome of both high-speed and luxury travel. They were introduced by the Italian state railways in the early 1950s. One "Settebello" still exists.

▷ BR Class AL1/ Class 81, 1959

Wheel arrangement Bo-Bo

Power supply 25 kV AC, overhead lines

Power rating 3,200 hp (2,387 kW)

Top speed 100 mph (161 km/h)

This was the first production AC electric locomotive class built in the UK for the first British 25 kV AC main line electrification of the London to Birmingham/Manchester/Liverpool line. As BR Class 81 the locomotives remained in service until 1991.

▽ DB Class E41/141, 1956

Wheel arrangement Bo-Bo

Power supply 15 kV AC, $16^2/_3$ Hz, overhead lines

Power rating 3,218 hp (2,401 kW)

Top speed 75 mph (120 km/h)

Large-scale plans for electrification of West Germany's railways during the 1950s led to large orders for several "Universal" locomotive types built by consortiums comprising all the major German locomotive-building firms. The E41 was the "universal" design for light passenger and freight trains. In total 451 were built between 1956 and 1971; all have now been withdrawn.

Post-war Steam

While railways played a vital strategic role in Europe during World War II, the ravages of war, destruction of industry, and shortages of raw materials and fuel painted a bleak picture for the Continent's future. Britain's railways and workshops escaped the worst excesses of destruction, and with innovative locomotive designers, such as Oliver Bulleid and Robert Riddles, were introducing new types of successful austerity locomotives towards the end of the war. In contrast, on mainland Europe the national railways were assisted in rebuilding their war-torn networks and rolling stock by deliveries of large numbers of powerful locomotives from US and Canadian manufacturers who were geared up to production through the Lend-Lease programme and the 1948 Marshall Plan.

△ SNCF 141R, 1945

Wheel arrangement	2-8-2
Cylinders	2
Boiler pressure	225 psi (15.82 kg/sq cm)
Driving wheel diameter	65 in (1,650 mm)
Top speed	approx. 62 mph (100 km/h)

Powerful and economical to maintain, 1,323 Class 141R locomotives were built between 1945 and 1947 for the French state railway (Société Nationale des Chemins de fer Français, or SNCF) by various builders in the US and Canada. Supplied under the Lend-Lease programme to replace engines lost during WWII, around half were oil-burners. Many remained in service until the 1970s.

△ Hunslet Austerity, 1944

Wheel arrangement	0-6-0ST
Cylinders	2 (inside)
Boiler pressure	170 psi (11.95 kg/sq cm)
Driving wheel diameter	51 in (1,295 mm)
Top speed	approx. 35 mph (56 km/h)

Designed by the Hunslet Engine Co. of Leeds, these locomotives were chosen by the British War Department for use as its standard shunting engine during WWII. Introduced in 1944, the earlier batches saw action in Europe and North Africa, as well as on military bases and ports across Britain.

▷ SNCB 29, 1945

Wheel arrangement	2-8-0
Cylinders	2
Boiler pressure	231 psi (16.24 kg/sq cm)
Driving wheel diameter	59 in (1,500 mm)
Top speed	approx. 60 mph (96 km/h)

After WWII these powerful mixed-traffic engines were built in Canada under the Lend-Lease programme to help in the reopening of Belgium's ruined state railways – Société Nationale des Chemins de fer Belges. Of the 180 built, one example, No. 29.013, has been preserved and is on display at the Belgian national railway museum at Schaarbeek.

◁ SR Bulleid Light Pacific, 1945

Wheel arrangement	4-6-2
Cylinders	3 (1 inside)
Boiler pressure	280 psi (19.68 kg/sq cm)
Driving wheel diameter	74 in (1,880 mm)
Top speed	approx. 90 mph (145 km/h)

Built under wartime conditions, Oliver Bulleid's "Battle of Britain" and "West Country" Class Light Pacific locomotives incorporated many cost-saving and innovative features. The 110 locomotives built for the Southern Railway and British Railways between 1945 and 1951 were renowned for their performance but suffered from high coal consumption. Sixty were subsequently rebuilt.

◁ GWR Modified Hall, 1944

Wheel arrangement 4-6-0

Cylinders 2

Boiler pressure 225 psi (15.82 kg sq cm)

Driving wheel diameter 70 in (1,778 mm)

Top speed approx. 75 mph (121 km/h)

Fitted with a large, three-row superheater to make up for the low-quality coal then available, these engines were a development by Frederick Hawksworth of Charles Collett's Hall Class. Between 1944 and 1950, a total of 71 were built at the Great Western Railway's Swindon Works.

◁ PKP Class Pt47, 1948

Wheel arrangement 2-8-2

Cylinders 2

Boiler pressure 213 psi (15 kg/sq cm)

Driving wheel diameter 72³⁄₄ in (1,850 mm)

Top speed approx. 68 mph (109 km/h)

Built by Fablok and Cegielski for the Polish state railways (Polskie Koleje Panstwowe, or PKP) from 1948 to 1951, these engines achieved outstanding performances hauling heavy passenger trains over long distances.

△ SNCF 241P, 1948

Wheel arrangement 4-8-2

Cylinders 4 (2 high-pressure, 2 low-pressure)

Boiler pressure 284 psi (19.96 kg/sq cm)

Driving wheel diameter 78³⁄₄ in (2,000 mm)

Top speed 75 mph (121 km/h)

These powerful "Mountain"-type express passenger compound locomotives were built by Schneider for the French state railway (SNCF) between 1948 and 1952. Designed to haul trains weighing 800 tons (813 tonnes) on the Paris to Marseilles main line, they were soon made redundant by electrification.

◁ Andrew Barclay Industrial, 1949

Wheel arrangement 0-4-0ST

Cylinders 2

Boiler pressure 160 psi (11.25 kg/sq cm)

Driving wheel diameter 35¹⁄₂ in (900 mm)

Top speed approx. 20 mph (32 km/h)

Scottish locomotive company Andrew Barclay built 100s of these diminutive saddle tanks for use on privately owned industrial railways in Britain and abroad. Their short wheelbase enabled them to operate on the sharply curved lines at collieries, steel and gas works, and docks.

N&W J Class No. 611

No. 611 is the sole remaining example of the Norfolk & Western (N&W) Railway's mighty J Class 4-8-4s, built at Roanoke, Virginia, between 1941 and 1950. With its streamlined front end, large cylinders, and roller-bearings all round, the locomotive was built for running in excess of 100 mph (161 km/h) and regularly plied the N&W routes from Cincinnati to Norfolk and Portsmouth. Today, No. 611 is preserved at the Virginia Museum of Transportation.

THE STORY OF the N&W's streamlined J Class 4-8-4s was in many respects defined by World War II. The engines were designed to haul the N&W's prestigious, named express trains, such as the *Powhatan Arrow* and the *Pocahontas*, with the first five (Nos. 600–604) completed in 1941–42. However, their introduction came just as the US entered the war, and this was reflected in the second batch (Nos. 605–610), which was delivered in 1943. Due to wartime material shortages, these six locomotives were constructed without streamlining and light-weight rods.

The final, streamlined, batch (Nos. 611–613) did not appear until 1950, but their career was short-lived. By the late 1950s the N&W had begun experimenting with diesel locomotives, and steam was displaced by the end of the decade. Thanks in part to the efforts of the American railway photographer O. Winston Link, No. 611 survived the cutter's torch and was donated to the Virginia Museum of Transportation, where it was returned to service in 1982.

FRONT VIEW

REAR VIEW

Fire Up 611
The builder's plate shows that 611 is a J Class completed in May 1950. Retired nine years later, 611 was eventually overhauled to pull excursion trains from 1982 until 1994. The "Fire Up 611" campaign is raising funds for a full restoration.

SPECIFICATIONS			
Class	J	In-service period	1950-59, 1982-94 (No. 611)
Wheel arrangement	4-8-4	Cylinders	2
Origin	USA	Boiler pressure	300 psi (21.09 kg/sq cm)
Designer/builder	Roanoke Shops	Driving wheel diameter	70 in (1,778 mm)
Number produced	14 J Class	Top speed	approx. 110 mph (177 km/h)

Tender carried by two 6-wheel bogies

"Tuscan red" stripe across full length of running board and tender

Firebox grate covers an area of 107 sq ft (10 sq m)

Roller bearings fitted to all crank pins and axles for smoother running

Streamlined casing with bullet nose

Black bullet
With its midnight-black livery, bullet-shaped nose, and powerful headlight, Norfolk & Western's No. 611 locomotive displays many of the characteristics so familiar to US streamliners.

EXTERIOR

At 109 ft (33 m) in length, 16 ft (4.9 m) in height, and weighing over 389 tons (392 tonnes), No. 611 is an impressive locomotive and was the pride of the N&W Railway. Its purposeful lines are accentuated by the heavy appearance of its coupling and connecting rods.

1. Number plate (one either side of headlight) **2.** Chrome strips at front of skyline casing **3.** Headlight **4.** Chrome marker lights **5.** Front steps to running board **6.** Control rod to throttle (regulator) **7.** Sander valve **8.** Handrail along running board **9.** Air compressor under front side of engine **10.** Lubrication system reservoir **11.** Sander **12.** Brake mechanism **13.** Driving wheels and connecting rods **14.** Injector **15.** Cab window **16.** Doors to tender coal bunker **17.** Stoker screw inside tender

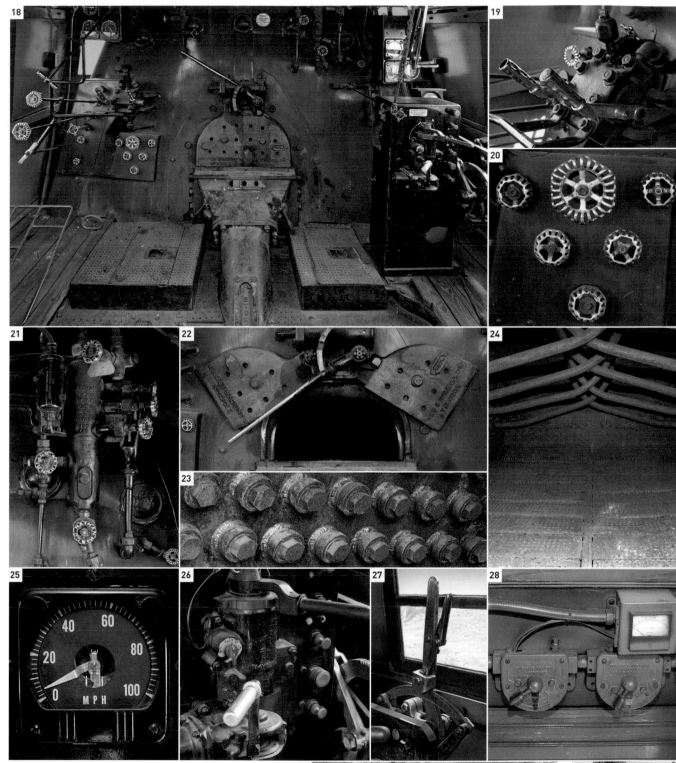

CAB INTERIOR

A locomotive of this size would be too much for a single fireman to manage using a shovel in the traditional style. Therefore, No. 611 was fitted with a mechanical stoker, which fed coal directly from the tender to the firebox by means of an Archimedes screw.

18. Cab interior **19.** Control levers for automatic grate shakers **20.** Control valves for stoker jets in firebox **21.** Gauge test valves and water level sight glass **22.** Open firebox door **23.** Staybolt detail inside firebox **24.** Circulators inside firebox **25.** Speedometer **26.** Brake control levers and handles **27.** Power reverse lever **28.** Electrical switches **29.** Throttle (regulator) quadrant **30.** Fireman's seat **31.** Foot rest

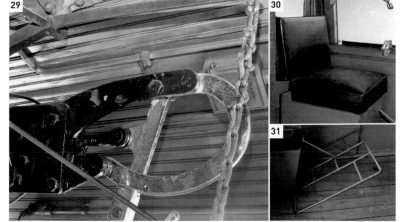

World Steam's Last Stand

With seemingly unlimited supplies of cheap foreign oil, by the 1960s many European and North American railways had replaced their steam engines with modern diesel-electric and electric engines, which were not only more efficient, powerful, and cleaner, but also required less maintenance between journeys. However, in other parts of the world where coal supplies were abundant and labour was cheap, steam continued to reign for a few more decades. In South Africa the development of steam locomotive design reached its pinnacle in the 1980s with the "Red Devil". Ending in 2005, the awesome spectacle of QJ 2-10-2 double-headed freight trains running through the frozen wastes of Inner Mongolia marked the final chapter of steam's 200-year reign.

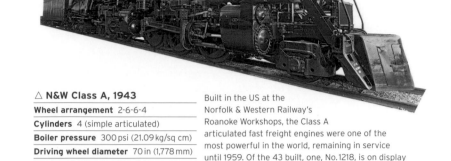

△ N&W Class A, 1943

Wheel arrangement	2-6-6-4
Cylinders	4 (simple articulated)
Boiler pressure	300 psi (21.09 kg/sq cm)
Driving wheel diameter	70 in (1,778 mm)
Top speed	70 mph (113 km/h)

Built in the US at the Norfolk & Western Railway's Roanoke Workshops, the Class A articulated fast freight engines were one of the most powerful in the world, remaining in service until 1959. Of the 43 built, one, No. 1218, is on display at the Virginia Museum of Transportation in Roanoke.

△ IR Class WP, 1947

Wheel arrangement	4-6-2
Cylinders	2
Boiler pressure	210 psi (14.78 kg/sq cm)
Driving wheel diameter	67 in (1,700 mm)
Top speed	68 mph (109 km/h)

Featuring a distinctive cone-shaped nose decorated with a silver star, 755 of the Class WP express passenger engines were built for the Indian broad-gauge railways between 1947 and 1967. No. 7161 *Akbar* is preserved at the Rewari Steam Loco Shed, India.

◁ Soviet Class P36, 1949

Wheel arrangement	4-8-4
Cylinders	2
Boiler pressure	213 psi (15 kg/sq cm)
Driving wheel diameter	73 in (1,854 mm)
Top speed	78 mph (126 km/h)

Built between 1949 and 1956, the 251 Class P36 were the last Soviet standard class, first working on the Moscow to Leningrad line until replaced by diesels. They later saw service in Eastern Siberia until being put into strategic storage from 1974 to the late 1980s.

▷ N&W J Class, 1950

Wheel arrangement	4-8-4
Cylinders	2
Boiler pressure	300 psi (21.09 kg/sq cm)
Driving wheel diameter	70 in (1,778 mm)
Top speed	70 mph (113 km/h)

A total of 14 J Class express passenger locomotives were built at the Norfolk & Western Railway's Roanoke Workshops between 1941 and 1950. Fitted with futuristic streamlined casings, they were soon replaced by diesels and had all retired by 1959.

▷ UP Class 4000 "Big Boy", 1941

Wheel arrangement	4-8-8-4
Cylinders	4
Boiler pressure	300 psi (21.09 kg/sq cm)
Driving wheel diameter	68 in (1,730 mm)
Top speed	80 mph (129 km/h)

Twenty-five of these monster articulated locomotives were built by the American Locomotive Co. (ALCO) for the Union Pacific Railroad between 1941 and 1944. Nicknamed "Big Boys", they were designed to haul heavy freight trains unaided over the Wasatch Range between Wyoming and Utah before being replaced by diesels in 1959. Eight have been preserved of which No. 4014 is being restored to working order.

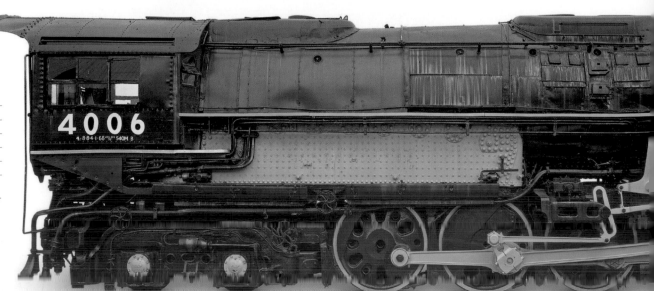

△ IR Class YG, 1949

Wheel arrangement 2-8-2

Cylinders 2

Boiler pressure 210 psi (14.8 kg/sq cm)

Driving wheel diameter 48 in (1,220 mm)

Top speed 50 mph (80 km/h)

The Class YG was the standard freight locomotive on the Indian Railways 3-ft 3-in- (1-m-) gauge system. Around 1,000 were built by various manufacturers in India and overseas between 1949 and 1972. Three, including *Sindh* seen here, are preserved in working order at Rewari Steam Loco Shed southwest of Delhi.

TALKING POINT

Cutting-edge Steam

Apart from the Class 25 condensing engines, South African Railways also took delivery of 50 Class 25NC (non-condensing). Of these, No. 3450 was modified in 1981 at the SAR's Salt River Workshops in Cape Town as the prototype Class 26. Nicknamed the "Red Devil" because of its livery, tests demonstrated vastly increased power and savings; diesel and electric traction had virtually replaced steam by the early 1980s.

The Red Devil This unique engine is seen here leaving Krankuil with a South African rail tour in 1990. It last ran in 2003 and is now preserved in Cape Town.

▷ SAR Class 25C, 1953

Wheel arrangement 4-8-4

Cylinders 2

Boiler pressure 225 psi (15.81 kg/sq cm)

Driving wheel diameter 60 in (1,524 mm)

Top speed 70 mph (113 km/h)

A total of 90 Class 25C locomotives were built for the 3-ft-6-in- (1.06-m-) gauge South African Railways. The engines were originally fitted with an enormous condensing tender so that they could operate across the arid Karoo Desert. Most were later converted to a non-condensing Class 25NC between 1973 and 1980.

△ IR Class WL, 1955

Wheel arrangement 4-6-2

Cylinders 2

Boiler pressure 210 psi (14.8 kg/sq cm)

Driving wheel diameter 67 in (1,702 mm)

Top speed 60 mph (96 km/h)

Featuring a light axle load for work on branch lines, these broad-gauge steam engines were built for the Indian Railways in two batches: the first 10 by Vulcan Foundry (UK) and 94 at the Chittaranjan Locomotive Works (India). No. 15005 *Sher-e-Punjab* is preserved at Rewari.

▷ China Railways CS Class QJ, 1956

Wheel arrangement 2-10-2

Cylinders 2

Boiler pressure 213 psi (15 kg/sq cm)

Driving wheel diameter 59 in (1,500 mm)

Top speed 50 mph (80 km/h)

One of the most prolific classes constructed in China was the Class QJ heavy freight engine of which at least 4,700 were built between 1956 and 1988. Their service on the Jitong Railway in Inner Mongolia (China) ended in 2005, though some ran on industrial railways until 2010.

Class WP No. 7161

Manufactured in the US by the Baldwin Locomotive Works, the first 16 Class WP steam locomotives – W for 5-ft 6-in- (1.67-m-) broad gauge and P for passenger – arrived in India in 1947. Chittaranjan Locomotive Works of West Bengal built No.7161 in 1965 to run on the Northeast Frontier Railway. Now based at Rewari Steam Loco Shed, where it was named *Akbar* after the Mughal emperor, it is the only working locomotive of its class.

WHEN THE CLASS WP of sleek, bullet-nosed, mainline steam locomotives was first introduced, it set the standard on Indian railways and became the mainstay of broad-gauge passenger operations for the rest of the 20th century. Known for free steaming, high fuel economy, and superior riding characteristics – and without the tail wag of the earlier X classes – its arrival marked the change of broad gauge coding from X to W.

Initially imported until 1959, the Class WP was manufactured in India between 1963 and 1967 at the Chittaranjan Locomotive Works, where 259 engines were built. Requiring a crew of three – a driver and two firemen – these locomotives hauled most of the prestigious passenger trains on the Indian railway system for the next 25 years. They established a sound reputation during their time in service, with their good performance earning them the title "Pride of the Fleet".

FRONT VIEW

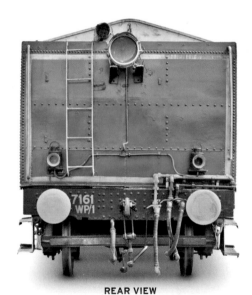

REAR VIEW

Heritage shed
Converted to a heritage museum by Indian Railways in 2002, the Rewari Steam Loco Shed houses some of India's last surviving steam locomotives.

SPECIFICATIONS			
Class	WP	**In-service period**	1965–96 (No. 7161)
Wheel arrangement	4-6-2	**Cylinders**	2
Origin	India	**Boiler pressure**	210 psi (14.78 kg/sq cm)
Designer/builder	Chittaranjan Locomotive Works	**Driving wheel diameter**	67 in (1,702 mm)
Number produced	755 (259 in India) Class WP	**Top speed**	68 mph (109 km/h)

Tender could carry 16 tons (16.2 tonnes) of coal and 6,500 gallons (29,550 litres) of water

Cab accommodates a crew of three

Air brake pipes outside locomotive frame

Metal chains along the length of the running board

Chimney is topped by a decorative crown

Bullet nose mounted on smokebox

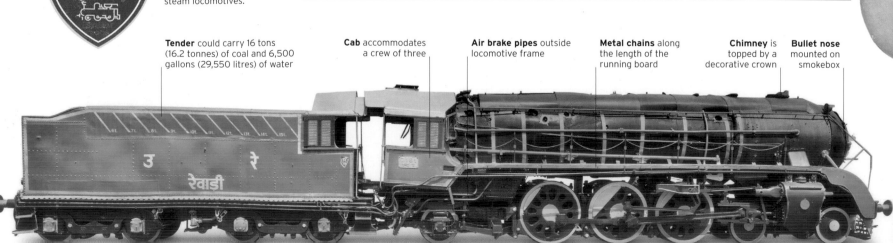

India's star locomotive
The decorative bullet nose bears a
silver star and is the most distinctive
feature of the locomotive. A nameplate
with the name *Akbar* sits centrally
below the nose.

EXTERIOR

With its distinctive bullet nose, a crown on top of the chimney, a 4-6-2 wheel arrangement, and a side profile that includes chain-decorated footboards along the length of the boiler, No.7161 is regarded as one of the most majestic locomotives that has ever run on Indian Railways. These features have proved popular with railway enthusiasts and tourists, and the locomotive currently hauls a mainline tourist train.

1. Hand-painted name on plaque at front **2.** Brass crown decorating top of chimney **3.** Headlight in the centre of a metal star **4.** Pilot light lamp, one positioned on either side of engine **5.** Cattle guard **6.** Steam chest valves **7.** Steam chest **8.** Driving wheels, with balance weight and connecting rod **9.** Big end and motion **10.** Rear carrying wheel **11.** Steps leading to cab **12.** Entrance to cab with wooden-slatted windows **13.** Light at back of tender **14.** Engine number **15.** Ladder at back of tender **16.** Rear buffer

CAB INTERIOR

The cab is spacious enough to house the driver and two firemen. The extra space also allows the firemen to stoke coal using shovels that are larger than those used in earlier locomotives. The red-painted handles for operating the locomotive and the monitoring gauges are positioned for ease of use.

17. Interior of driver's cab **18.** Lubricator box **19.** From left to right: injector steam cock handle, dynamo cock, main cock, vacuum steam cock, and injector steam cock handle **20.** Steam pressure gauge **21.** Reverser wheel **22.** Firehole door **23.** Rocking grate **24.** Front of tender

Europe's Last Gasp

With diesel and electric traction rapidly gaining favour, the 1950s saw the last steam locomotives built for Europe's national railways. In West Germany the last one to be built for Deutsche Bundesbahn, No. 23.105, rolled off the production line in 1959. Across the English Channel, Robert Riddles had designed 12 new classes of standard locomotives for the nationalized British Railways. Sadly, many of these fine engines had extremely short working lives owing to the hurried implementation of the ill-conceived Modernisation Plan. Despite this, privately owned British locomotive manufacturers such as Beyer Peacock & Co. of Manchester and Hunslet of Leeds continued to export steam locomotives; the last engine was built by Hunslet in 1971.

◁ DB Class 23, 1950

Wheel arrangement	2-6-2
Cylinders	2
Boiler pressure	232 psi (16.3 kg/sq cm)
Driving wheel diameter	69 in (1,750 mm)
Top speed	68 mph (110 km/h)

This engine was designed to replace the Prussian Class P8 passenger locomotives on the West German Deutsche Bundesbahn. The 105 Class 23s were built between 1950 and 1959. No. 23.105 was the last steam locomotive built for DB. The final examples were retired in 1976 and eight have been preserved.

△ *Bonnie Prince Charlie*, 1951

Wheel arrangement	0-4-0ST
Cylinders	2
Boiler pressure	160 psi (11.25 kg/sq cm)
Driving wheel diameter	24 in (610 mm)
Top speed	20 mph (32 km/h)

Built by Robert Stephenson & Hawthorns in 1951, *Bonnie Prince Charlie* originally worked as a gas works shunter at Hamworthy Quay in Dorset (UK). It was bought by the Salisbury Steam Trust in 1969 and has since been restored at Didcot Railway Centre.

▷ BR Class 4MT, 1951

Wheel arrangement	2-6-4T
Cylinders	2
Boiler pressure	225 psi (15.82 kg/sq cm)
Driving wheel diameter	68 in (1,730 mm)
Top speed	70 mph (113 km/h)

Robert Riddles's Class 4MT tank locomotive was the largest of four standard tank designs built by British Railways. Used primarily on suburban commuter services, a total of 155 were built between 1951 and 1956 but were soon displaced by electrification.

▷ BR Class 9F, 1954

Wheel arrangement	2-10-0
Cylinders	2
Boiler pressure	250 psi (17.57 kg/sq cm)
Driving wheel diameter	60 in (1,524 mm)
Top speed	90 mph (145 km/h)

The Class 9F was the standard heavy freight locomotive built by British Railways between 1954 and 1960. A total of 251 were built with No. 92220 *Evening Star* being the last steam engine built for BR. Although designed for freight haulage, they were occasionally used on express passenger duties. All were retired by 1968, and nine have been preserved.

▷ **BR Class 7 Britannia, 1951**

Wheel arrangement	4-6-2
Cylinders	2
Boiler pressure	250 psi (17.57 kg/sq cm)
Driving wheel diameter	74 in (1,880 mm)
Top speed	90 mph (145 km/h)

A total of 55 Class 7 Britannia engines, designed by Robert Riddles, were built at British Railway's Crewe Works between 1951 and 1954. After hauling expresses across the BR network, they were relegated to more humble duties. One lasted until the end of BR mainline steam in 1968.

△ **DR Class 65.10, 1954**

Wheel arrangement	2-8-4T
Cylinders	2
Boiler pressure	232 psi (16.3 kg/sq cm)
Driving wheel diameter	63 in (1,600 mm)
Top speed	56 mph (90 km/h)

The powerful Class 65.10 tank locomotives were built to haul double-deck and push-pull commuter trains on the Deutsche Reichsbahn in East Germany. All 88 built had retired by 1977, but three have been preserved.

◁ **DR Class 99.23-24, 1954**

Wheel arrangement	2-10-2T
Cylinders	2
Boiler pressure	203 psi (14.27 kg/sq cm)
Driving wheel diameter	39½ in (1,003 mm)
Top speed	25 mph (40 km/h)

Seventeen of these massive 3-ft 3-in- (1-m-) gauge tank locomotives were built for the Deutsche Reichsbahn in East Germany from 1954 to 1956. They still survive on the highly scenic railways in the Harz Mountains with nine currently in working order.

△ **Beyer-Garratt Class NG G16, 1958**

Wheel arrangement	2-6-2+2-6-2
Cylinders	4
Boiler pressure	180 psi (12.65 kg/sq cm)
Driving wheel diameter	33 in (840 mm)
Top speed	40 mph (64 km/h)

Several European manufacturers built 34 of these engines from 1937 to 1968 for the 2-ft- (0.61-m-) gauge lines of South African Railways. No.138, built by Beyer Peacock & Co., now hauls trains on the Welsh Highland Railway.

Beyer-Garratt No.138

The NG G16 class of Beyer-Garratts has achieved international prominence serving the Welsh Highland Railway, but these locomotives were originally built for mining concerns in southern Africa. Used mainly for freight in Africa, in Wales these engines have a new life pulling passenger carriages in Snowdonia National Park, showing off their haulage capacity and articulation on the steep gradients and sharp curves of the mountainous landscape.

BEYER-GARRATT NO.138 is one of the last batch of Garratt locomotives built by Beyer, Peacock & Company Ltd in Manchester. The Tsumeb Corporation of South West Africa (now Namibia) ordered seven locomotives of this type to haul minerals from its mines in the Otavi mountains. However, the re-gauging of the 256-mile- (412-km-) long Otavi Railway to 3 ft 6 in (106 cm) before the locomotives arrived resulted in their sale to South African Railways for use in Natal, on the east coast.

Allocated to the 76-mile- (122-km-) long Port Shepstone-Harding line, No.138 was one of the assets transferred when the railway was privatized in 1986. The locomotive was withdrawn from service in 1991. However, after being selected by the Ffestiniog Railway for use on the Welsh Highland Railway (WHR) in 1993, it was overhauled at Port Shepstone and delivered to Wales. It began running on the WHR in green livery in October 1997, but was painted red in 2010.

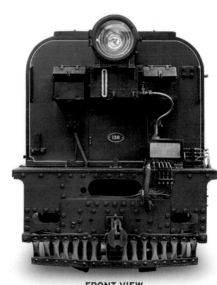

FRONT VIEW

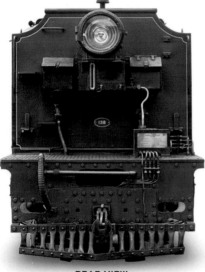

REAR VIEW

Preserving history
At 25 miles (40 km), the Welsh Highland Railway is the longest heritage railway in Britain. The railway stopped running services before WWII but a restoration project was completed in 2011.

SPECIFICATIONS			
Class	NG G16	In-service period	1958-91 and 1997 to present (No.138)
Wheel arrangement	2-6-2+2-6-2	Cylinders	4
Origin	UK	Boiler pressure	180 psi (12.65 kg/sq cm)
Designer/builder	Beyer, Peacock & Co. Ltd	Driving wheel diameter	33 in (840 mm)
Number produced	34 Class NG G16	Top speed	approx. 40 mph (64 km/h)

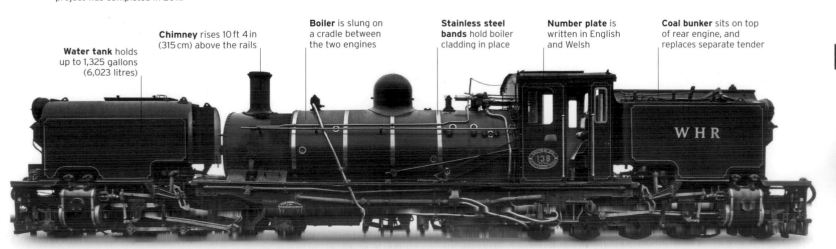

Water tank holds up to 1,325 gallons (6,023 litres)

Chimney rises 10 ft 4 in (315 cm) above the rails

Boiler is slung on a cradle between the two engines

Stainless steel bands hold boiler cladding in place

Number plate is written in English and Welsh

Coal bunker sits on top of rear engine, and replaces separate tender

Great power
The NG G16 Garratt is the largest and most powerful narrow-gauge steam locomotive in Britain. It weighs 62 tons (63 tonnes) and is 46 ft 6 in (14.7 m) long. Each end of the locomotive is equipped with powerful headlamps, sand boxes, and mechanical lubricators.

EXTERIOR

Garratt locomotives consist of three main components – two engines and a boiler cradle. The cradle is pivoted to the engines on both ends to provide the articulation that enables the locomotive to traverse sharp curves. The additional wheel sets provided by the duplicated engine reduce the weight carried by each axle, so that it can operate on lighter rail. As a result, NG G16 locomotives such as No. 138 can run safely on rails as light as 40 lb per yard (20 kg per metre), although Welsh Highland Railway rail weighs 60 lb per yard (30 kg per metre). The locomotives were designed to be operated equally well in either direction.

1. Numberplate **2.** Level indicator of lubricator oil reservoir **3.** Headlamp **4.** Lubricator **5.** Water tank filler cover **6.** Coupler **7.** Leaf spring suspension **8.** Die block **9.** Washout plug **10.** Top clack valve **11.** Dome cover **12.** Chime whistle **13.** Crosshead and cylinder **14.** Water filter **15.** Coal bunker

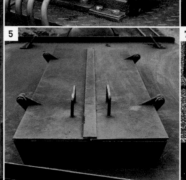

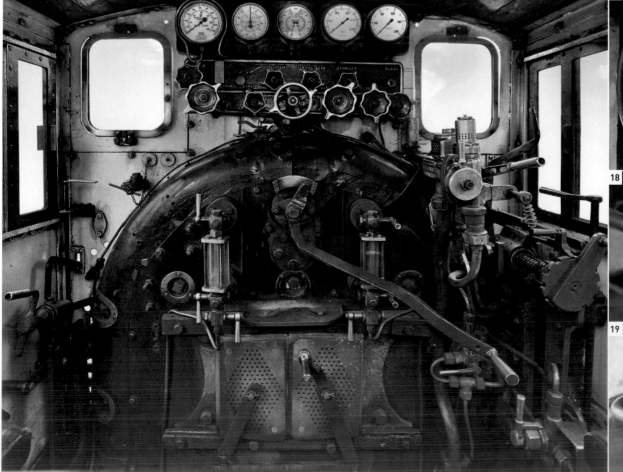

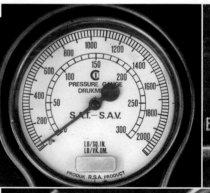

CAB INTERIOR

Despite the complexity of its mechanical arrangements, No. 138's cab is much like that of any steam locomotive. The driver's controls are on the right-hand side. As well as looking after the fire to create steam, the fireman is responsible for operating the injectors, which put water into the boiler as and when required.

16. Controls in driver's cab **17.** Boiler pressure gauge **18.** Injector steam valve (left) and main manifold valve (right) **19.** Boiler pressure gauge isolator **20.** Cylinder drain, sander, and atomizer controls
21. Water gauge **22.** Reverser **23.** Vacuum brake controls
24. Speedometer **25.** Driver's seat

Moving People and Goods

Although the very first railways were built to carry freight, some were designed from the start primarily to transport passengers. During the world wars, the railways carried huge quantities of raw materials, military supplies, and troops. However, by the 1950s they struggled against more flexible and cheaper road transport. Meeting this challenge with some success, the railways carved out the vital roles of transporting commuters. They competed with air travel by introducing faster and more luxurious passenger trains and focussed on the long-haul, heavy-freight traffic that remains a core business today.

△ **GWR Corridor Composite carriage No. 7313, 1940**

Type 2 x 4-wheel bogies

Capacity 24 first-class passengers plus 24 third-class passengers

Construction steel

Railway Great Western Railway

Built by the Great Western Railway at their Swindon Works in 1940, the 60-ft-(18.2-m-) long express passenger coach No. 7313 has four first-Class compartments, four third-Class compartments, and two lavatory cubicles. It is wearing its "wartime economy" brown livery and is preserved at Didcot Railway Centre.

△ **N&W Budd S1 sleeper, 1949**

Type 2 x 4-wheel bogies

Capacity 22-32 sleeping berths

Construction stainless steel

Railway Norfolk & Western Railway

Twenty of these sleeping cars were built by Budd in 1949 for the Norfolk & Western Railway. They were used on the *Powhatan Arrow*, *The Pochohontas*, and other sleeping car routes on the railway's network. *The Pochohontas*, the N&W's last passenger train, ceased running in 1971. This car is now preserved at the Virginia Museum of Transport in Roanoke.

▷ **N&W Pullman Class P2 No. 512, 1949**

Type 2 x 4-wheel bogies

Capacity 66 passengers

Construction steel

Railway Norfolk & Western Railway

Seating 66 passengers, this coach was built for the Norfolk & Western Railway's *Powhatan Arrow* by Pullman-Standard in 1949. Introduced between Norfolk, Virginia, and Cincinnati, Ohio, in 1946, the train last ran in 1969. This coach is now on display at the Virginia Museum of Transport in Roanoke.

Freight Cars

Road transport began siphoning off much of the peacetime short-distance, single-load freight traffic, but the railways' trump card was their ability to transport heavy loads more efficiently over long distances. To meet this demand a wide variety of purpose-built freight cars were constructed to carry raw materials such as coal, oil, and iron ore; perishable goods such as fish, meat, fruit and vegetables; and hazardous cargoes such as chemicals and petroleum.

△ **Penn Central Wagon No. 32367, 1955**

Type Class H34A covered hopper

Weight 62½ tons (63.5 tonnes)

Construction steel

Railway Penn Central

The Wagon No. 32367 was built at the Penn Central Corporation's Altoona Workshops in 1955. The cargo (often grain) was discharged through chutes underneath the wagon. It is now on display at the Railroad Museum of Pennsylvania in Strasburg.

▷ **VEB double-deck coach, 1951**

Type	2- to 5-car articulated coach sets
Capacity	approx. 135 passengers per coach
Construction	steel
Railway	Deutsche Reichsbahn

Known as Doppelstockwagen in Germany, these double-deck coaches are descended from those introduced on the Lübeck–Büchen Railway in 1935. Built by Waggonbau Görlitz, they were capable of carrying 50 per cent more passengers than single-deck coaches. Seen here are the first of around 4,000 double-deck, articulated coaches built in East Germany on a test run in 1951.

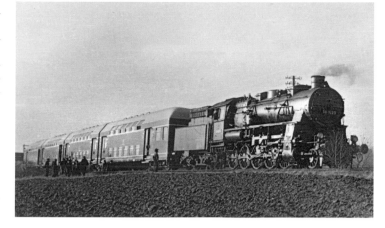

Travelling in Comfort

While the Railway Regulations Act of 1847 made it compulsory for Britain's railways to provide poorer people with travelling accommodation at an affordable price, the well-heeled traveller was charged much more for comfort. Up until 1956 there were three classes of travel – first, second, and third. Second class was then abolished. First-class compartments offered plenty of legroom, and luxury seating, carpets, and curtains. Third-class passengers were squashed into more basic compartments with horse-hair seats.

Class distinction The first-class compartment (below, left) features curtains, carpets, and individual wingbacks and armrests for its six passengers. Third class (below, right) has a less comfortable bench-seat arrangement.

◁ **BR(W) Brake Third carriage No. 2202, 1950**

Type	2 x 4-wheel bogies
Capacity	24 Third Class passengers plus guard's and luggage compartments
Construction	steel
Railway	British Railways (Western Region)

Featuring distinctive domed roof ends and designed by the Great Western Railway's last chief mechanical engineer, F. W. Hawksworth, this Brake Third carriage was built in 1950 for British Railways (Western Region) by Metropolitan-Cammell of Birmingham. It is now preserved at Didcot Railway Centre.

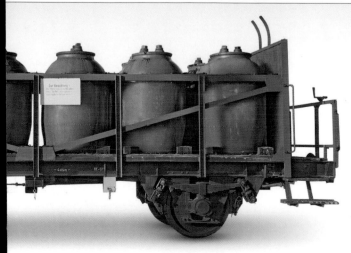

◁ **DR Acid Cannister Wagon, 1956**

Type	cannister wagon
Weight	14.6 tons (14.83 tonnes)
Construction	steel
Railway	Deutsche Reichsbahn

Built in 1956 for the East German state railways, this freight wagon carried 12 clay pots, each containing 220 gallons (1,000 litres) of acid. It is on display at the Stassfurt Museum Shed.

△ **MDT/IC No. 13715, 1958**

Type	refrigerated boxcar
Weight	37½ tons (38 tonnes)
Construction	steel
Railway	Illinois Central Railroad

This 33-ft- (10-m-) long, insulated, refrigerated boxcar was built by the Pacific Car & Foundry Co. of Renton in Washington State for the Illinois Central Railroad in 1958. Fitted with air circulation fans, this type of car usually carried perishable fruit and vegetables, which were kept chilled by dry ice loaded into roof-mounted bunkers.

1960–1979
BUILT FOR SPEED

Glasgow Electric

Travel by the Modern Railway

BUILT FOR SPEED

When Japan's first Shinkansen railway opened in 1964, it heralded an exciting future for rail. With its special high-speed lines and modern electric units, the "Bullet Train" revolutionized the way passengers experienced rail travel. Japan offered an exciting vision of the future, and railways in the West were inspired to innovate. Streamlining and modernizing, operators introduced new diesel and electric trains, refurbished stations, built new freight facilities, invested in infrastructure, and continued to increase the speed on existing lines. In some nations, this was the era when steam locomotives were finally retired from service. "Inter-city" travel became the norm.

However, this fresh emphasis on speed was not enough to revitalize railway travel to the level of its heyday. The popularity of train travel began to decline with the rise in car ownership and an increase in jetliner travel, which became more widely available. As a result, many rural and other less profitable lines were closed. In some countries the proposals were drastic – Britain's Beeching Report, published in 1963, recommended the closure of around 30 per cent of the network. In the US a government-backed organization, Amtrak, was formed in 1971 with a responsibility for rescuing the unprofitable, long-distance passenger services.

The situation in Eastern Europe was different. The absence of mass car ownership ensured that passenger demand for rail travel remained high; railways were considered strategically vital too. Modernization in this region often meant increasing train capacity, as opposed to cutting lines, and speeds remained relatively low on the whole. Elsewhere, however, by the mid-1970s many countries had started to follow Japan's lead, creating their own high-speed trains.

> " There's a great **emotional upsurge** every time we intend to **cancel a service**"
>
> DR RICHARD BEECHING,
> CHAIRMAN OF BRITISH RAILWAYS

△ **Amtrak Turboliner**
The modern, fast Turboliner was introduced by Amtrak in 1973 in an effort to encourage more passenger rail travel.

◁ **Glasgow Electric poster** by the English painter Terence Tenison Cuneo, 1965

Key Events

▷ **1960** British Railways follows the global trend and stops building steam locomotives. The last one, a freight engine, is named *Evening Star*.

▷ **1961** The building of the Berlin Wall forces a revamp of rail services to and from the western parts of the city.

▷ **1963** The Beeching Report heralds a drastic downsizing of Britain's railways.

▷ **1964** The launch of Shinkansen train services in Japan pioneers a new form of high-speed rail transport.

△ **Launch of the "Bullet Train"**
The opening of the Tōkaidō Shinkansen line was accompanied by an official ceremony at Japan National Railway's Tokyo station on 1 October 1964.

▷ **1971** Amtrak is formed to rescue inter-city rail travel in the US, after private companies find passenger trains increasingly unviable.

▷ **1972** France's experimental gas-turbine TGV 001 is finished. It takes the world rail speed record by reaching 198 mph (318 km/h).

▷ **1973** Britain's High Speed Train (HST) prototype achieves a diesel world record – about 143 mph (230 km/h).

▷ **1974** The USSR makes completion of the Baikal-Amur Magistral a national priority, to provide a second route to complement the Trans-Siberian.

▷ **1976** Work starts on France's first dedicated high-speed line, to run between Paris and Lyon. It is the beginning of the country's dedicated high-speed network.

Freight and Passenger Accelerates

During the 1960s and 1970s railways around the world followed the early lead of North America and replaced steam with either diesel or electric locomotives. The growth in car ownership in many Western countries meant that railways had to offer faster and more comfortable trains to persuade passengers to use the train instead. Freight services – historically very slow – gathered speed through the introduction of new locomotives that were twice as fast and twice as powerful as the steam locomotives they replaced.

△ BR Type 4 Class 47, 1962

Wheel arrangement Co-Co

Transmission electric

Engine Sulzer 12LDA28-C

Total power output 2,750 hp (2,051 kW)

Top speed 95 mph (153 km/h)

The most numerous main-line diesel locomotives ever used in the UK, the first 20 Class 47s were delivered in 1962/63 and tested on British Railway's Eastern Region. Orders for more soon followed, and a total of 512 were built by both Brush Traction's Falcon Works and BR's Crewe Works. Some remain in use with British operators.

▷ Soviet Class M62, 1964

Wheel arrangement Co-Co

Transmission electric

Engine Kolomna V12 14D40

Total power output 1,973 hp (1,472 kW)

Top speed 62 mph (100 km/h)

The Soviet M62 design was exported to Warsaw Pact countries in the 1960s and 1970s, as well as being delivered to Soviet Railways. Between 1966 and 1979 Czechoslovakia received 599 of them from Voroshilovgrad Locomotive works (in present-day Ukraine). Production only ended in 1994 and one is shown here.

△ DR V180, 1960

Wheel arrangement B-B

Transmission hydraulic

Engine 2 x 12KVD21 A-2

Total power output 1,800 hp (1,342 kW)

Top speed 75 mph (120 km/h)

The V180 was designed to replace steam engines on main-line passenger and freight trains in two versions – as well as the initial 87 four-axle versions, a further 206 more powerful six-axle locomotives, were delivered by 1970 and subsequently renumbered as DR Class 118.

◁ DR V100, 1966

Wheel arrangement B-B

Transmission hydraulic

Engine MWJ 12 KVD 18-21 A-3

Total power output 987 hp (736 kW)

Top speed 50 mph (80 km/h)

The East German V100 centre-cab design was first tested in 1964, and in total 1,146 production locomotives of several types were built for the Deutsche Reichsbahn from 1966 to 1985. The V100s were also exported to several other communist countries such as Czechoslovakia and China.

▽ GM EMD Class SD45, 1965

Wheel arrangement Co-Co

Transmission electric

Engine 20-cylinder EMD 645E3

Total power output 3,600 hp (2,685 kW)

Top speed 65 mph (105 km/h)

General Motors Electro-Motive Division (EMD) built 1,260 SD45 locomotives from 1965 to 1971 for several US railways, using a 20-cylinder version of EMD's then new 645 engine. Some SD45s remain in use in the US freight railroads. Shown here is Erie Lackawanna Railway's No. 3607, which has been preserved.

△ GM EMD GP40, 1965

Wheel arrangement	Bo-Bo
Transmission	electric
Engine	16-645E3
Total power output	3,000 hp (2,237 kW)
Top speed	65 mph (105 km/h)

Baltimore & Ohio Railroad bought 380 General Motors Electro-Motive Division (EMD) model GP40 locomotives so had the largest fleet in the US of these successful locomotives. In total 1,221 were built for various operators in North America between 1965 and 1971. They were used for freight trains by B&O but other operators used them for passenger services.

◁ DB Class 218 (V160), 1971

Wheel arrangement	B-B
Transmission	hydraulic
Engine	MTU MA 12 V 956 TB 10
Total power output	2,467 hp (1,840 kW)
Top speed	87 mph (140 km/h)

The Deutsche Bundesbahn first ordered the final version of the V160 fleet – Class 218 – in the late 1950s. The prototypes were delivered in 1968 and 1969; series production began in 1971. Fitted with electric train heating, the Class 218 could work with the latest air-conditioned passenger coaches. Of the 418 delivered, around half remain in use.

△ Chinese DF4, 1969

Wheel arrangement	Co-Co
Transmission	electric
Engine	16V240ZJA
Total power output	3,251 hp (2,425 kW)
Top speed	62 mph (100 km/h)

The DF4, known as "Dong Feng" (East Wind), is one of a series of locomotives built for the Chinese national railways. Updated versions remain in production over 40 years after the first one was built at China's Dalian Locomotive Works. DF4s replaced steam locomotives throughout China and several thousand remain in use.

TECHNOLOGY

Container Transport

The use of containers to transport freight by ship began in the 1950s. In 1952 Canadian Pacific introduced the "piggyback" transport of containers on wheeled road trailers, although the Chicago North Western Railroad had pioneered this before World War II. During the 1960s rail operators started to offer services to transport the maritime containers (called "intermodal" as they can be transferred from one form of transport to another) to and from ports on specially designed flat wagons. Intermodal freight transport grew substantially in the 1970s and 1980s. In 1957 it accounted for less than one per cent of US rail freight, but by the mid 1980s more than 15 per cent of freight was transported in this way.

B&O Class P-34 No. 9523 This is a 40-ton (40.64-tonne) flat car for carrying road semitrailers. It was built by B&O in 1960 at its workshops in Dubois, Pennsylvania.

Modified DR V100

The East German V100 diesel hydraulic was first tested in 1964, and eventually 1,146 production locomotives of several versions were built for the Deutsche Reichsbahn (DR) between 1966 and 1985. They were also made for heavy industry, and exported to several other communist countries such as Czechoslovakia and China. In 1988, just before the fall of the Berlin Wall (in 1989), conversion of 10 locomotives for metre-gauge operation began.

THE DEUTSCHE REICHSBAHN ORDERED several versions of the V100 type to replace steam engines on local passenger and freight trains, and to be used for heavy shunting. They were built by East Germany's VEB Lokomotivbau Elektrotechnische Werke "Hans Beimler" Hennigsdorf (LEW), which occupied the site of AEG's pre-war Hennigsdorf factory, north of Berlin.

From 1988, 10 locomotives were converted for the 3-ft 3-in- (1-m-) gauge network in the Harz Mountains of central Germany, where they gained the nickname "Harzkamel" (Harz camel). They were intended to be the first of 30 locomotives to replace steam but, after the Harz system was privatized as a network focusing on tourism, steam engines were retained for most trains, and there was little work for the "Harz camels". Two were converted to work freight trains where standard-gauge wagons were carried on new 3-ft 3-in- (1-m-) gauge transporter bogies. Several have now been sold and converted back to standard gauge, and along with many other DR V100s remain in use with freight operators in Germany and elsewhere.

FRONT VIEW

REAR VIEW

SPECIFICATIONS			
Class	HSB 199.8 – previously V100, then DR 112, DB 202	**In-service period**	1966-78, as rebuilt 1988-present (No. 119 872-3)
Wheel arrangement	C-C, built as B-B	**Transmission**	hydraulic
Origin	East Germany	**Engine**	MWJ 12 KVD 18-21 A-4
Designer/builder	LEW (Berlin),	**Power output**	1,184 hp (883 kW)
Number produced	10 (rebuilt 199.8 series)	**Top speed**	31 mph (50 km/h) (rebuilt 199.8 series)

Ventilation grilles for the engine

Exhaust takes fumes above top of cab

Driving cab in centre gives excellent visibility in all directions

Orange warning light used when locomotive is remotely controlled

Three axle bogies fitted at conversion to 3-ft 3-in (1-m-) gauge

Antenna for locomotive telecommunication system

HSB
199 872 -3

Harz network
The HSB logo stands for Harzer Schmalspurbahnen (Harz Narrow Gauge Railways), the operator of the Harz Mountain 3-ft 3-in (1-m-) gauge network since 1993.

Snow camels
The "Harzkamel" nickname came from the locomotive's waggling gait and the camel "hump" formed by the central cab. These "kamels", however, were more at home in mountains than deserts, and the plough used to clear a small coverage of snow can be seen below the buffers at track level.

EXTERIOR

The body comprises two bonnets extending from the central cab. At one end is the engine, and at the other a variety of ancillary equipment such as steam heating equipment for passenger coaches and batteries. The hydraulic transmission system is located under the driving cab alongside the diesel fuel tank.

The locomotives were built with two-axle, standard-gauge bogies; in the conversion to 3-ft 3-in- (1-m-) gauge these were replaced with three-axle bogies utilizing smaller diameter wheels. The remaining HSB locomotives were rebuilt again in 1998, and three were fitted with GPS equipment enabling them to be controlled by yard staff.

1. Number plate **2.** Headlight (below) and tail light (above) **3.** Buffer in raised position **4.** Coupling for standard-gauge wagons **5.** Electric socket for multiple control unit **6.** Coupling for wagon carrier bogies **7.** Air brake pipe connecting adapter **8.** Open sandbox door **9.** Fuel filler **10.** Warning light used when remote controlled **11.** Overhead electrification warning flash **12.** Air horn **13.** Foot step to reach top of locomotive **14.** Cooling device for air compressors **15.** Filter, drain cup, and drip cock in main air pipe **16.** Wheel assembly **17.** Air shut off valves **18.** Socket for charging cable **19.** Grease container for flange oilers **20.** Steps for shunters **21.** Cut-off cock for main brake pipe

CAB INTERIOR

The cab, although spartan by modern standards, was functional and a lot simpler and cleaner than the steam engine cabs it replaced. It was designed to enable the locomotive to be driven in either direction. As was common in the Eastern Bloc, many components were interchangeable with other types to reduce the number of spare parts required.

22. Overview of cab interior **23.** Control lamps **24.** Cab controls **25.** Joystick for driving **26.** Timetable holder **27.** Speedometer **28.** Pressure gauge for brake cylinder and main brake pipe **29.** Handle for sliding cab window **30.** Dead-man's vigilance device, which checks that driver is not incapacitated **31.** Air valve for radiator **32.** Handle for cab window lock **33.** Light fitting

High-speed Pioneers

High-speed rail travel began in 1960 when French Railways introduced the world's first 124-mph (200-km/h) passenger train – the "Le Capitole" Paris to Toulouse service. In 1964 the first Japanese Shinkansen line from Tokyo to Shin-Osaka was opened; this was the start of fast passenger train services on a dedicated high-speed rail line. Higher-speed operations began in the UK with the 100-mph (161-km/h) "Deltic" diesels in 1961, and in North America with gas-turbine–powered trains in 1968. In the 1970s the German Class E03/103 began a 124-mph (200-km/h) operation on existing lines in West Germany, while in the UK the new diesel-powered High Speed Train (HST) brought 125-mph (201-km/h) services to several major routes from 1976.

△ DR Class VT18.16 (Class 175), 1964

Wheel arrangement	4-car DMU
Transmission	hydraulic
Engine	2 x 12 KVD 18/21 engines
Total power output	1,973 hp (1,472 kW)
Top speed	100 mph (160 km/h)

Built by East German industry to operate the Deutsche Reichsbahn's important international express trains, eight four-car VT18.16 trains were delivered from 1964 to 1968. These worked abroad reaching Copenhagen, Denmark; Vienna, Austria; and Malmö, Sweden; plus Prague and Karlovy Vary in Czechoslovakia. The trains were progressively withdrawn in the 1980s, although more than one survives.

▽ BR Type 5 *Deltic* D9000 Class 55, 1961

Wheel arrangement	CoCo
Transmission	electric
Engine	2 x Napier Deltic 18-25 engines
Total power output	3,299 hp (2,461 kW)
Top speed	100 mph (161 km/h)

Based on the *Deltic* prototype of 1955, a total of 22 of these engines were ordered for express passenger trains on British Railways' East Coast main line between London, York, Newcastle, and Edinburgh to replace 55 steam locomotives. Capable of sustained 100 mph (161 km/h) running, the class enabled faster trains to be operated on the route from 1963. Withdrawn in 1981, several have been preserved in working order.

△ JNR Shinkansen Series 0, 1964

Wheel arrangement	12-car EMU, all 48 axles powered
Power supply	25 kV AC overhead lines
Power rating	11,903 hp (8,880 kW)
Top speed	137 mph (220 km/h)

Japan built brand-new, standard-gauge (4-ft 8½-in/1.4-m) high-speed lines to dramatically improve journey times. The first section of Japan National Railways' Tōkaidō Shinkansen line operated at 130 mph (209 km/h) – at the time the fastest trains in the world.

TECHNOLOGY

Amtrak Begins Service

The US National Railroad Passenger Corporation (Amtrak) took over long-distance passenger rail services in May 1971, following a US Congress decision to maintain some level of rail service after many companies had moved to freight only. Amtrak started life with old equipment, but quickly started looking for new diesel and electric trains including new French-built Turboliner trains.

The turbo train Amtrak introduced six 125-mph (201-km/h) Turboliner trains from 1973 on services from Chicago. Powered by Turbomeca gas turbines originally designed for helicopters, the trains never got to exploit their high-speed capability.

△ DB Class E03/103, 1970

Wheel arrangement CoCo

Power supply 15 kV AC, 16⅔ Hz overhead lines

Power rating 10,429 hp (7,780 kW)

Top speed 124 mph (200 km/h)

Five E03 prototypes were delivered from 1965, and after test, another 145 slightly more powerful production engines were ordered. From 1970 until the 1980s the Deutsche Bundesbahn Class 103 worked on all the major express trains in Germany. A small number remain in use; one was used for high-speed test trains until 2013 and allowed to run at 174 mph (280 km/h).

▽ SNCF Class CC6500, 1969

Wheel arrangement CoCo

Power supply 1.5 kV DC overhead lines (21 locos also equipped for 1.5 kV DC third rail)

Power rating 7,909 hp (5,900 kW)

Top speed 124 mph (200 km/h)

Seventy-four powerful CC6500 engines were delivered between 1969 and 1975 to run on the Société Nationale des Chemins de fer Français' "Le Capitole" Paris to Toulouse service. Twenty-one were fitted with third-rail pick-up and pantographs, for use on the Chambéry-Modane "Maurienne" line.

△ UAC Turbo Train, 1968

Wheel arrangement 7-car articulated train set

Transmission torque coupler

Engine 4 x Pratt & Whitney Canada ST6B gas turbines

Total power output 1,600 hp (1,193 kW)

Top speed 120 mph (193 km/h)

United Aircraft Corporation (UAC) entered the market with patents bought from the Chesapeake & Ohio Railway for articulated high-speed train sets using lightweight materials. However, UAC used gas turbines instead of diesel engines. Canadian National Rail bought five sets and the US bought three.

▽ BR HST Class 253/254, 1976

Wheel arrangement BoBo

Transmission electric

Engine (power car) Paxman Valenta 12R200L

Total power output (power car) 2,249 hp (1,678 kW)

Top speed 125 mph (201 km/h)

In 1973 British Rail started trials of the High Speed Train prototype with two power cars. Production trains followed in 1976, with deliveries lasting until 1982. The HST holds the world diesel rail speed record of 148 mph (238 km/h) set in 1987. The trains remain in service as do similar ones in Australia.

The Bullet Train

The staging of the 1964 Summer Olympics in Tokyo presented Japan with the opportunity to show how far it had progressed since the devastation of World War II. The nation decided to showcase its engineering capabilities with the Tōkaidō Shinkansen, the world's first high-speed railway.

Construction of the electrified line, which ran 321.6 miles (515.4 km) and linked Tokyo with Osaka to the southwest, began in 1959 and was completed in 1964. Service commenced on 1 October that year. The line carried the world's fastest trains, which earned the nickname *Dangan Ressha* ("Bullet Trains") because of their speed and the distinctive shape of the leading car. Reaching a top speed of 130 mph (210 km/h), the

trains soon made the journey in a record-breaking 3 hours and 10 minutes. Popular from the outset, at peak times the service ran at 3-minute intervals.

The first line carried more than 150 million passengers in its inaugural year. Its success led to more routes on the islands of Honshu and Kyushu, enlarging the network to 1,483.6 miles (2,387.7 km). Engineers also designed faster models and tracks; even the original 0-series trains were modified, reaching a top speed of 200 mph (320 km/h) before they were retired in 2008.

The 16-car 300 series Shinkansen entered service in 1992, performing at a top speed of 168 mph (270 km/h). The series was taken out of service in 2012.

DR No. 18.201

East Germany's Deutsche Reichsbahn (DR) No.18.201 is one of a kind. Built to a unique design to allow the testing of coaches at high speeds, it is the world's fastest operational steam locomotive. Oil firing, a special streamlined casing, and massive driving wheels all helped to create a machine not only able to reach high speeds, but also to maintain them. Just as remarkably, it was built when steam development was all but over.

A SPECIAL SET of circumstances led to the creation of No. 18.201. East Germany required a method to test passenger coaches it was building for export, and felt that the most practical way to achieve this was to construct a high-speed steam locomotive fit for that purpose. To build the specialist machine, engineers used parts from older locomotives, including the high-speed tank engine No. 61.002 (which the DR had inherited after World War II), as well as new components.

The most recognizable parts taken from other locomotives were the goods engine tender and No. 61.002's big driving wheels. However, No. 18.201's streamlined look was distinctively modern. Unusual fittings included brakes on the locomotive's leading bogie, which gave it extra stopping power at high speeds; it also received the "Indusi" safety gear, designed to stop trains passing any stop signals. For most of its career No. 18.201 was based at the railway test facility in Halle (Saale) in Saxony-Anhalt. It is now cared for by the Dampf-Plus company at Lutherstadt Wittenberg.

FRONT VIEW

REAR VIEW

Deutsche Reichsbahn

Separate German networks
After World War II, West Germany's railway became the Deutsche Bundesbahn, but East Germany's system kept the traditional Deutsche Reichsbahn name. The two merged into the Deutsche Bahn in January 1994.

SPECIFICATIONS				
Class	18.2	In-service period	1961-present
Wheel arrangement	4-6-2	Cylinders	3
Origin	East Germany	Boiler pressure	232 psi (16.3 kg/sq cm)
Designer/builder	Deutsche Reichsbahn	Driving wheel diameter	91 in (2,311 mm)
Number produced	1	Top speed	approx. 113 mph (182 km/h)

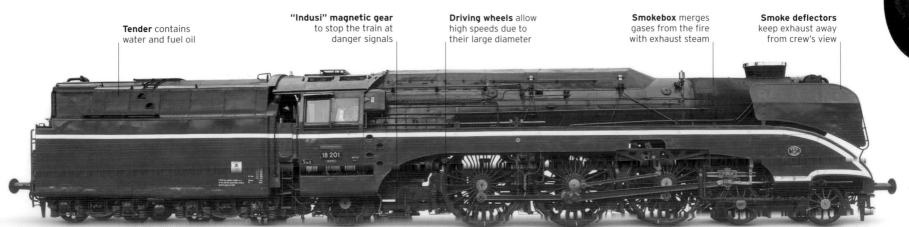

Tender contains water and fuel oil

"Indusi" magnetic gear to stop the train at danger signals

Driving wheels allow high speeds due to their large diameter

Smokebox merges gases from the fire with exhaust steam

Smoke deflectors keep exhaust away from crew's view

A touch of style
The curves and angles of No.18.201 gave the locomotive a stylishly modern look. The smokebox door was a distinctive conical shape, while the locomotive's thin but efficient *Giesl* ejector exhaust was hidden inside a much larger chimney shroud.

EXTERIOR

Although unique and instantly recognizable, No. 18.201 shares design elements with other German steam locomotives, as well as high-speed engines from elsewhere. The look is dominated by the green semi-streamlined casing and the large 91-in (2,311-mm) driving wheels, which allow the locomotive to run faster. Some details, such as the front headlamps, are non-standard add-ons, while others came from the Deutsche Reichsbahn stores.

1. Number plate on side of cab **2.** Front headlamp **3.** Coupling hook **4.** Front buffer **5.** Front steps **6.** Steam-powered electrical generator **7.** Shut-off valve **8.** Whistle **9.** Lagged pipework **10.** Valve gear **11.** Small End **12.** Bogie wheel **13.** Inside the Big End **14.** Lead driving wheel **15.** Sand pipe **16.** Brake assembly **17.** Air pump assembly **18.** Steps to cab at front of tender **19.** "Indusi" magnet **20.** Detail of tender bogie **21.** Headlamp on rear of tender **22.** Oil filler

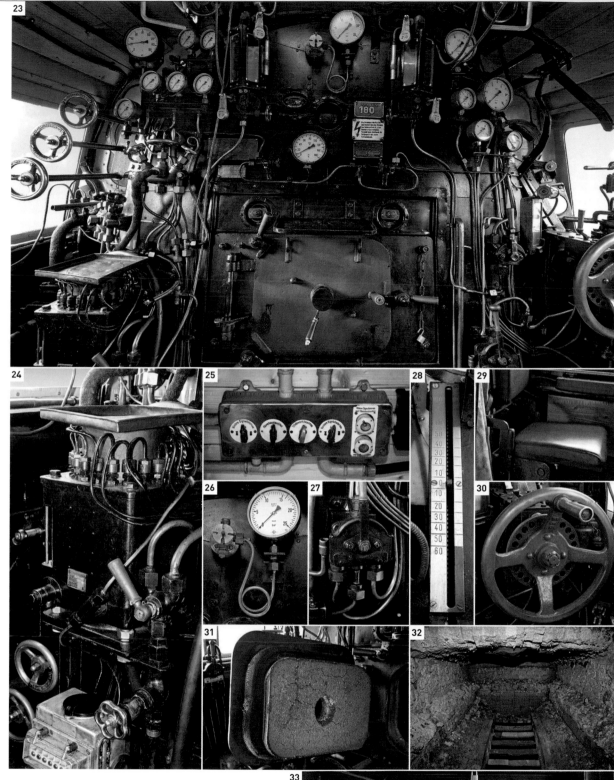

CAB INTERIOR

In comparison with coal-fired locomotives, the oil-fired No. 18.201 has different controls for the fireman to regulate the fire, as well as dials to monitor it. The insulated firebox door stays shut while the engine is operating, and the onerous task of shovelling coal is unnecessary. The driver sits on the right, where all the main driving controls are within easy reach.

23. Overview of cab controls **24.** Lubricator **25.** Lamp switches **26.** Pressure gauge **27.** Sanding controls
28. Reverse/cut-off indicator **29.** Fireman's seat
30. Reverser **31.** Firebox door **32.** Interior of firebox
33. Area at front of tender

Technology in Transition

This was a period of large-scale changes for railways around the world. Car ownership and the impact of new motorways led to the closure of less-used railway routes, particularly in Western Europe, although branch lines still thrived in Eastern Europe. Commodities such as coal and iron ore continued to be carried by the railways. Much of the local goods transport switched to trucks, but the use of intermodal containers to carry long-distance freight by rail continued to grow. In addition, many European cities were expanding existing metro systems or building new ones.

△ **BR D9500 Class 14, 1964**

Wheel arrangement	0-6-0
Transmission	hydraulic
Engine	Paxman 6YJXL
Total power output	650 hp (485 kW)
Top speed	40 mph (64 km/h)

The 56 locomotives in this class, all built at the British Railways Works in Swindon during 1964, were delivered just as the local freight traffic they were designed for was rapidly disappearing from the UK rail network. As a result many were withdrawn within three years. Most went on to have longer careers with industrial rail operators in the UK and Europe.

◁ **Soviet Class VL10, 1963**

Wheel arrangement	Bo-Bo+Bo-Bo
Power supply	3,000 V DC, overhead lines
Power rating	6,166 hp (4,600 kW)
Top speed	62 mph (100 km/h)

Built in Tbilisi (now Georgia) the VL10 eight-axle, twin-unit electric was the first modern DC electric locomotive built for Soviet Railways. It shared both external design and many components with the VL80 25 kV AC electric design, also introduced in 1963. Thousands of both classes were built until production ended in the 1980s.

△ **DR VT2.09 (Class 171/172), 1962**

Wheel arrangement	2-axle rail bus
Transmission	mechanical/hydro-mechanical
Engine	6 KVD 18 HRW
Total power output	180 hp (134 kW)
Top speed	56 mph (90 km/h)

Designed for East Germany's rural branch lines, this train was nicknamed "Ferkeltaxi" (piglet taxi) because farmers sometimes brought piglets along as luggage. An early prototype built in 1957 was followed by orders for production trains, delivered from 1962 to 1969. In 2004 they were withdrawn from regular use in Germany.

TALKING POINT

Track Maintenance

Motorized draisines replaced or supplemented daily track inspections carried out on foot from the 1960s, enabling tools and equipment to be carried to work sites quickly. During the 1960s ultrasonic testing of rails by test vehicles fitted with special equipment became more common in both US and Europe, and regular test trains operated, often at night, to monitor track condition.

Room for two In East Germany the two-axle draisine could carry two people and their tools to repair minor faults.

▷ **DR V60 D (Class 105), 1961**

Wheel arrangement	0-8-0
Transmission	hydraulic
Engine	12 KVD 18/21
Total power output	650 hp (485 kW)
Top speed	37 mph (60 km/h)

The powerful V60 was designed to replace the Deutsche Reichsbahn steam locomotives for shunting and short freight trains. The engines, enhanced by advances made on the WWII V36 diesels used by the German military, unusually had four axles with the wheels connected by external coupling rods. They were built for the DR and other state railways plus heavy military purposes.

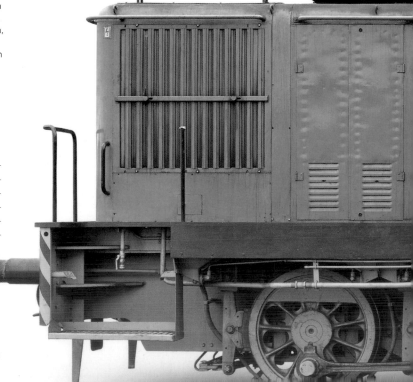

TECHNOLOGY

Battery Locomotives

In many European countries battery-powered engines were used to move locomotives around maintenance depots. Using the battery engines enabled electric locomotives to be transferred to maintenance areas without (hazardous) overhead power lines for traction current and was quicker and cheaper than starting a diesel to move it a few hundred yards. Battery engines continue to be used in this way today.

Akkuschleppfahrzeuge (ASF) Over 500 ASFs (meaning battery-shunting vehicle) were built in East Germany from 1966 to 1990. Used by DR and industrial operators, some are still working.

△ Preston Docks Sentinel, 1968

Wheel arrangement A-A

Transmission hydraulic

Engine Rolls-Royce C8SFL

Total power output 325 hp (242 kW)

Top speed 18 mph (29 km/h)

The Sentinel locomotives were designed to replace steam engines at major industrial sites that operated their own railways. Innovative and easy to use, they had a central driving position in a full-width cab and safe places for shunting staff to travel on the outside of the engines. Several are preserved at UK heritage railways.

△ LT Victoria Line, 1969

Wheel arrangement 4-car units, always operated as pairs

Power supply 630 V DC third and fourth rail system

Power rating 1,137 hp (848 kW)

Top speed 25 mph (40 km/h)

The Victoria Line was the first completely new Tube line in London for 60 years when it opened in 1969. The new trains bought by London Transport were fitted with Automatic Train Operation (ATO) equipment – the train drove itself and the "driver" would normally only open and close doors at stations.

▽ DR V300 (Class 132), 1973

Wheel arrangement Co-Co

Transmission electric

Engine Kolomna 5D49

Total power output 3,000 hp (2,237 kW)

Top speed 74 mph (120 km/h)

Based on the Soviet TE109 design and built at Voroshilovgrad (now Luhansk, Ukraine), the most numerous of the DR V300 locomotives was Class 132, with 709 locomotives. While most have been withdrawn, some remain in service with several German freight operators today.

Great Journeys
Indian Pacific

The first direct passenger rail service to cross the continent of Australia from the east coast to the west, the *Indian Pacific* finally linked Sydney on the Pacific Ocean to Perth on the Indian Ocean on 23 February 1970.

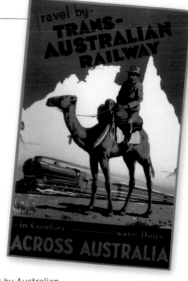

AUSTRALIA'S FIRST UNBROKEN transcontinental railway was made possible only by standardizing the random mixture of broad-, standard-, and narrow-gauge lines that had been built in the 19th and early 20th centuries.

The New South Wales government opened the state's first standard-gauge railway in 1855, linking Sydney on the east coast to nearby Granville. This track was gradually extended over the Blue Mountains via a series of steeply graded zigzags, reaching Orange – 200 miles (322km) from Sydney – in 1877. From Orange, the standard-gauge Broken Hill line opened westwards in stages, between 1885 and 1927, across sparsely populated, arid lands to the mining town of Broken Hill. Westwards from Broken Hill, the 3-ft 6-in- (1.06-m-) gauge Silverton Tramway, opened in 1888, reached as

Indian Pacific stops in Broken Hill
A 4,000-hp (2,984-kW) NR Class diesel-electric locomotive pulling the *Indian Pacific* halts at Broken Hill. The town is at the centre of the world's largest silver, lead, and zinc ore deposits.

Saving days by train
A vintage travel poster by Australian artist James Northfield publicizes the advantages of the newly built Trans-Australian section of the railway.

far as Cockburn. Here the railway met the South Australian Railways' line of the same gauge from Port Pirie, part of the Adelaide to Port Augusta line.

From the west coast, a 3-ft 6-in- (1.06-m-) gauge line already linked Perth to the gold-mining town of Kalgoorlie by 1897. Between Kalgoorlie and Port Augusta remained a 1,000-mile (1,609-km) gap across South Australia through a region that was a virtually uninhabited and waterless desert.

In 1901 the newly formed Commonwealth of Australia's government proposed a railway to link isolated Western Australia with the rest of the country. Opened throughout in 1917, the 1,052-mile (1,693-km) Trans-Australian Railway across the aptly named Nullarbor ("no tree") Plain was built to the standard gauge of 4ft 8½in (1.435m), but met with narrow-gauge lines at either end. No natural

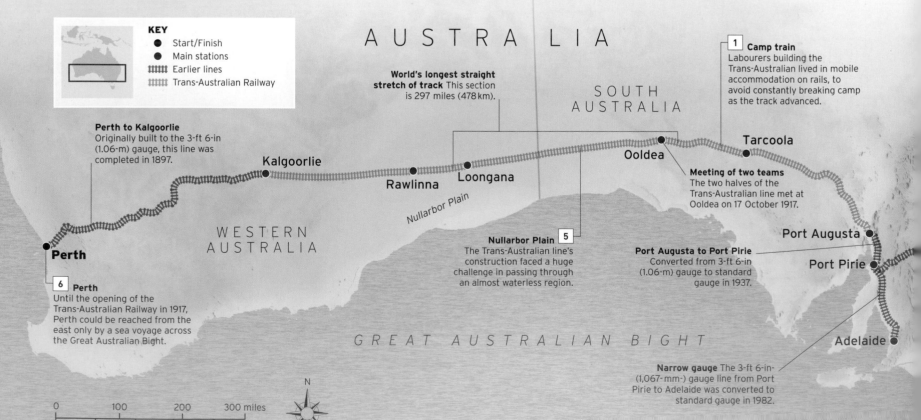

KEY
- ● Start/Finish
- ● Main stations
- ⊪⊪⊪ Earlier lines
- ⊪⊪⊪ Trans-Australian Railway

AUSTRALIA

World's longest straight stretch of track This section is 297 miles (478km).

1 Camp train
Labourers building the Trans-Australian lived in mobile accommodation on rails, to avoid constantly breaking camp as the track advanced.

SOUTH AUSTRALIA

Perth to Kalgoorlie
Originally built to the 3-ft 6-in (1.06-m) gauge, this line was completed in 1897.

Kalgoorlie

Rawlinna

Loongana

Nullarbor Plain

WESTERN AUSTRALIA

Ooldea

Tarcoola

Meeting of two teams
The two halves of the Trans-Australian line met at Ooldea on 17 October 1917.

Nullarbor Plain 5
The Trans-Australian line's construction faced a huge challenge in passing through an almost waterless region.

Port Augusta to Port Pirie
Converted from 3-ft 6-in (1.06-m) gauge to standard gauge in 1937.

Port Augusta

6 Perth
Until the opening of the Trans-Australian Railway in 1917, Perth could be reached from the east only by a sea voyage across the Great Australian Bight.

Perth

Port Pirie

GREAT AUSTRALIAN BIGHT

Adelaide

Narrow gauge The 3-ft 6-in- (1,067-mm-) gauge line from Port Pirie to Adelaide was converted to standard gauge in 1982.

0   100   200   300 miles

N

water sources existed on this stretch of the line, so steam-hauled trains had to carry their own supplies, which occupied over half the train's load. Diesels took over in 1951.

A unified standard-gauge railway across the continent was realized in stages: the line from Port Augusta to Port Pirie was converted in 1937, and the 374-mile (602-km) line from Perth to Kalgoorlie was converted in 1969. The track between Port Pirie and Broken Hill was rebuilt as standard gauge by 1970, and the *Indian Pacific* made its first run from Sydney. Now a luxury train, it completes the four-day journey twice-weekly, stopping off at the historic Broken Hill and offering an experience of remote Australian terrain.

In 1982 the *Indian Pacific* also began to call at Adelaide after the line south of Port Pirie was converted to standard gauge, extending the distance travelled by the train to 2,704 miles (4,352 km).

Across the Nullarbor
Double-headed by NR Class diesels, the *Indian Pacific* heads out across the arid Nullarbor Plain on the world's longest straight stretch of track.

KEY FACTS

DATES
1917 Standard-gauge Trans-Australian Railway completed, meeting existing narrow-gauge lines in east and west.
1970 Continuous standard-gauge railway between Sydney and Perth completed. *Indian Pacific* inaugural run on 23 February.

TRAIN
First locomotives Commonwealth Railways CL Class 3,000 hp (2,238 kW) Co-Co diesel-electrics built 1970-72
Current locomotives NR class 4,000 hp (2,984 kW) Co-Co diesel-electrics built 1996-98
Carriages Up to 25 75-ft (23-m) air-conditioned stainless steel carriages, including sleeping cars, restaurant car, power van, luggage van, and Motorail wagons carrying passengers' cars. Three classes: Platinum, Gold, and Red

JOURNEY
Original journey: Sydney-Perth 2,461 miles (3,961 km); 75 hours
Sydney-Perth (via Adelaide) 2,704 miles (4,352 km), 65 hours; 4 days, 3 nights

RAILWAY
Gauge Standard gauge 4 ft 8 ½ in (1.435 m)
Longest straight stretch The world's longest section of straight track, 297 miles (478 km)
Highest point Bell Railway Station in the Blue Mountains: 3,507 ft (1,069 m)

ACROSS A CONTINENT

On its 65-hour journey across New South Wales, South Australia, and Western Australia, the *Indian Pacific* crosses three time zones, which were introduced in the 1890s. Perth is two hours behind Sydney in summer, and three in winter.

QUEENSLAND

4 Mannahill Station
Located along the *Indian Pacific* route, Mannahill is one of the easternmost settlements in South Australia. It has only 66 inhabitants.

NEW SOUTH WALES

Broken Hill

Blue Mountains The rail routes originally crossed the mountains on steeply graded zigzags which were bypassed in 1910.

Cockburn Ivanhoe Condobolin

Orange

Granville

3 Sydney
Australia's premier east coast city is famous for its iconic Opera House and Harbour Bridge.

AUSTRALIAN CAPITAL TERRITORY

Sydney

PACIFIC OCEAN

Broken Hill to Port Pirie
Converted from 3-ft 6-in (1.06-m) gauge to standard gauge in 1970.

VICTORIA

2 Inaugural journey
The *Indian Pacific*, the first train to cross the entire Australian continent, left Sydney on 23 February 1970.

Travelling in Style

In the 1960s and 70s railways around the world invested in large numbers of new passenger carriages. The investment was partly driven by the need to offer higher speed and more comfort on intercity routes, and in other cases simply to replace older equipment. Steel became the dominant material for coach bodies, replacing wooden-framed, steam-age vehicles in many cases. Increasing numbers of new multiple-unit trains, both diesel and electric, were built in many countries to replace conventional trains using locomotives and coaches.

◁ Cravens Stock, 1963

Type	second-class, open coach
Capacity	64 passengers
Construction	steel
Railway	CIÉ (Irish railways)

Fifty-eight of these coaches were assembled in Irish Railways's Inchicore Works in Dublin between 1963 and 1967, using kits provided by Cravens in Sheffield, UK. The coaches were fitted with steam heating and vacuum brakes, and were used for express trains in the 1960s. Several coaches have been preserved.

△ Talgo III, 1964

Type	articulated express passenger car
Capacity	21 passengers
Construction	stainless steel
Railway	RENFE (Spanish state railways)

In the 1950s the Spanish Talgo company pioneered articulated trains of semi-permanently coupled short cars utilizing single-axle wheel sets. The Talgo III was the third version of the train and the first to be used internationally. Some had variable-gauge axles, which permitted operation from Spain into France.

△ Penn Central/Amtrak Metroliner, 1969

Type	snack bar car (powered)
Wheel arrangement	2-car EMU
Power supply	11 kV AC 25 Hz, 11 kV AC 60 Hz, and 25 kV AC 60 Hz, overhead lines
Power rating	1,020 hp (761 kW)
Top speed	125 mph (200 km/h)

Budd built 61 Metroliner EMU cars for Penn Central Transportation in 1969 in collaboration with other manufacturers and the US government. The cars were inherited by Amtrak in 1971. Designed for use at 150 mph (241 km/h), the Metroliners never operated that fast and most were withdrawn by Amtrak in the 1980s.

△ Reko-Wagen, 1967

Type	second-class, open coach
Capacity	64 passengers
Construction	steel
Railway	Deutsche Reichsbahn

The Deutsche Reichsbahn introduced *Reko-Wagen* (reconstructed coaches) in the 1950s and 60s – the reconstruction referred to their rebuild from older designs. Initially, short three-axle coaches were built but in 1967 ㎝ㅑ ㎢ ㎗ long bogie coaches appeared.

△ Eurofima, 1973

Type	first- and second-class open
Capacity	54 (first); 66 (second)
Construction	steel
Railway	SBB (Swiss Railways; and others)

In the mid-1970s several Western European railways jointly ordered 500 new daytime coaches to a standard design following tests with 10 prototypes. They were funded via Eurofima, a not-for-profit rail financing organization based in Switzerland. In total 500 coaches were built for six different operators.

▽ **Mark IIIB First Open, 1975**

Type first class Pullman coach

Capacity 48 passengers

Construction steel

Railway British Rail

The first 125 mph (201 km/h) Mark III coaches appeared in 1975 and incorporated steel integral monocoque construction, giving them great body strength. The British Rail High Speed Train (HST) used Mark III coaches and others were built for use with electric locomotives at up to 110 mph (177 km/h).

△ **Mark III sleeper, 1979**

Type sleeping coach

Capacity 26 berths in 13 compartments

Construction steel

Railway British Rail

In 1976 British Rail ordered a new prototype sleeper with a view to replacing its older cars, but this was cancelled after a fatal fire on Mark I sleepers on an overnight train in Taunton in 1978. BR decided to build a new version that incorporated safety systems onto all sleepers; 236 were ordered in 1979.

▷ **Amtrak Superliner, 1978**

Type double-deck long distance

Capacity up to 74, fewer for sleepers

Construction stainless steel

Railway Amtrak

Based upon cars originally built in 1956 for the Atchison, Topeka & Santa Fe Railway and inherited by Amtrak in 1971, the Superliner long-distance cars were built from 1978. Nearly 500 were made over the next 20 years in multiple configurations (sleepers, seating cars, diners, and observation cars).

Danger
Overhead
live wires

Class	92		A
Weight tonnes			126
Brake force tonnes			53
ETH index			:
ac			180
dc			108
RA			7
Max speed km/h			140
Braked Weight			:
Goods tonnes			65
Passenger tonnes			94

Cab 1
Side B

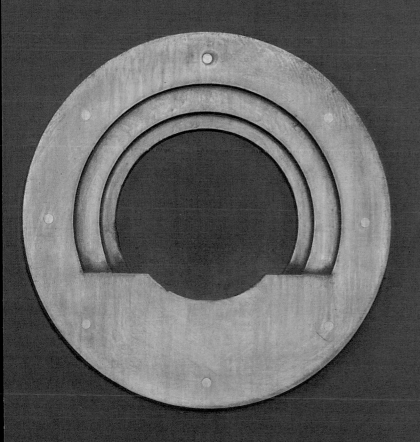
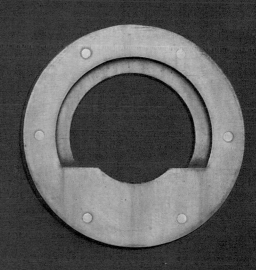

CHANGING TRACKS

CHANGING TRACKS

The high-speed railway spread internationally as more countries built dedicated networks replicating the Japanese invention. In Europe, France's Train à Grande Vitesse (TGV) was launched in 1981 with a line running from Paris to Lyon, and a decade later Germany saw the InterCityExpress (ICE) make its public debut. In the UK, however, the emphasis lay on modernizing the existing system, rather than building new lines.

As the renaissance in light rail continued, new tram systems opened in some places. Karlsruhe in southwest Germany introduced a new concept: the "tram-train", a vehicle capable of running both on the streets and on local railways. Yet while rail technology improved, there was also a desire for "golden age" travel inspired by the past, which was realized with the launch of classic luxury trains such as Europe's *Venice Simplon-Orient Express* and India's *Palace on Wheels*.

The end of the Cold War ushered in changes to Europe's railways, not least in Germany. Following the country's unification in 1990, lines that ran across the former border were reopened and new ones were built, and the former East and West German systems were eventually merged as the Deutsche Bahn.

However, the restructuring was much more radical in the UK after the British parliament voted for privatization in 1993. In the years that followed, the state-owned British Rail was dismantled; new companies took over different lines and implemented their own plans for development, and in doing so reintroduced variety to the train services.

In 1994 rail celebrated yet another engineering marvel with the opening of the Channel Tunnel, which connected France and the UK for the first time. Running under the Dover Strait, the launch of the tunnel was the realization of a dream dating back to the 19th century.

"So speed yes, but let there be money in it"

GERARD FIENNES,
FIENNES ON RAILS, 1986

△ **Across the Channel**
On 14 November 1994, Eurostar services began between London Waterloo International, Paris Gare du Nord, and Brussels-South.

◁ **A Union Pacific freight train** winding across the US landscape

Key Events

▷ **1981** High-speed rail services come to Europe when France launches the Train à Grande Vitesse (TGV). It raises the world speed record to 236 mph (380 km/h).

△ **High-speed rail in France**
The TGV-PSE is a high-speed train built for operation between Paris and the southeast of France. The original fleet had an orange and silver livery.

▷ **1991** Russia completes the Baikal-Amur Magistral – a major main line paralleling the classic Trans-Siberian.

▷ **1991** Germany enters the public high-speed rail era with the InterCityExpress (ICE).

▷ **1992** "Tram-train" services are launched in Karlsruhe, Germany. The new concept unites local rail and tramways with vehicles that can run on both systems.

▷ **1993** Britain votes to privatize its railways. In the years that follow, the state system is split up.

▷ **1994** In Germany, the former West German Deutsche Bundesbahn and East German Deutsche Reichsbahn are merged to form a new entity – the Deutsche Bahn.

▷ **1994** The Channel Tunnel opens, connecting Britain and France by rail underneath the Dover Strait.

▷ **1995** China's Ji-Tong Railway opens in Inner Mongolia. Known as the world's last steam main line, it is not fully converted to diesel until 2005.

High Speed Goes Global

Operating at speeds that were impossible on historic railway tracks, high-speed lines had burst upon the world scene in 1964 with the introduction of the Shinkansen in Japan. In Europe the French led the way, building a network of dedicated high-speed lines known as a Train à Grandes Vitesse (TGV), with the first route between Paris and Lyon opening in 1981. Spain's first high-speed line, the Alta Velocidad Española (AVE), opened between Madrid and Seville in 1992. The UK, with its Victorian rail network, lagged behind; despite the opening of the Channel Tunnel in 1994, it would not be until 2007 before the country's first dedicated high-speed railway HS1 was complete, ushering in high-speed rail travel between London and Paris.

▷ Soviet ER200, 1984

Wheel arrangement each car
2 x 4-wheel bogies

Power supply 3 kV DC overhead lines

Power rating 6-car set: 5,150 hp
(3,840 kW)/14-car set: 15,448 hp
(11,520 kW)

Top speed 124 mph (200 km/h)

Built of aluminium alloy in Riga, the ER200 is a Soviet high-speed train that was first introduced in 1984. At the time it was the first Direct Current (DC) intercity electric multiple-unit train with rheostatic braking. Later versions operate on the Moscow to St Petersburg main line. Unit ER200-15 is on display at the Moscow Railway Museum.

△ AVE S-100, 1992

Wheel arrangement each car
2 x 4-wheel bogies

Power supply 3 kV DC overhead
supply/25 kV 50 Hz AC overhead supply

Power rating 11,796 hp (8,800 kW)

Top speed 186 mph (300 km/h)

The Alta Velocidad Española (AVE) is a network of high-speed railways operated in Spain by Renfe Operadora. It was Europe's longest high-speed network and, after China, the world's second longest. The first line between Madrid and Seville opened in 1992 using S-100 dual-voltage, electric multiple units built by Alstom.

▷ Thalys PBKA, 1996

Wheel arrangement 2 power cars
+ 8 passenger cars

Power supply 3 kV DC overhead supply/
25 kV 50 Hz AC overhead supply/15 kV
16$\frac{2}{3}$ Hz AC overhead supply/1,500 V DC
overhead supply

Power rating 4,933 hp (3,680 kW) –
11,796 hp (8,800 kW)

Top speed 186 mph (300 km/h)

Built by GEC-Alstom in France, the Thalys PBKA is a high-speed international train service, introduced in 1996, that can operate on four different electrical systems in France, Germany, Switzerland, Belgium, and the Netherlands. The 17 train sets built operate services between Paris, Brussels, Cologne (Köln), and Amsterdam, hence PBKA.

TECHNOLOGY

Transrapid Prototype

Developed in Germany, this high-speed monorail train with no wheels, gear transmissions, or axles, and has no rails or overhead power supply. Instead it levitates, or hovers, above a track guideway using attractive magnetic force between two linear arrays of electromagnetic coils, hence its name "Maglev". Based on a patent from 1934, planning for it began in 1969 and the test facility was completed in 1987. The latest version Maglev 09 can cruise at over 300 mph (482 km/h). The only commercial application to date opened in China in 2002 and operates between Shanghai and its Pudong international airport.

Revolutionary technology The two-car Maglev Transrapid prototype is seen in action at the test facility at the Emsland test track in Germany in 1980.

▷ Eurostar Class 373/1, 1993

Wheel arrangement each car
2 x 4-wheel bogies

Power supply 25 kV 50 Hz AC overhead
supply/3,000 V DC overhead supply/1,500 V DC
overhead supply/750 V DC third-rail (not used)

Power rating 4,600 hp (3,432 kW) –
16,360 hp (12,200 kW)

Top speed 186 mph (300 km/h)

Introduced in 1993, the Class 373/1 multi-voltage electric multiple units are operated by Eurostar on the high-speed line between London, Paris, and Brussels via the Channel Tunnel. In the UK these trains operated on the third-rail network to London's Waterloo Station until the completion of the HS1 line in 2007.

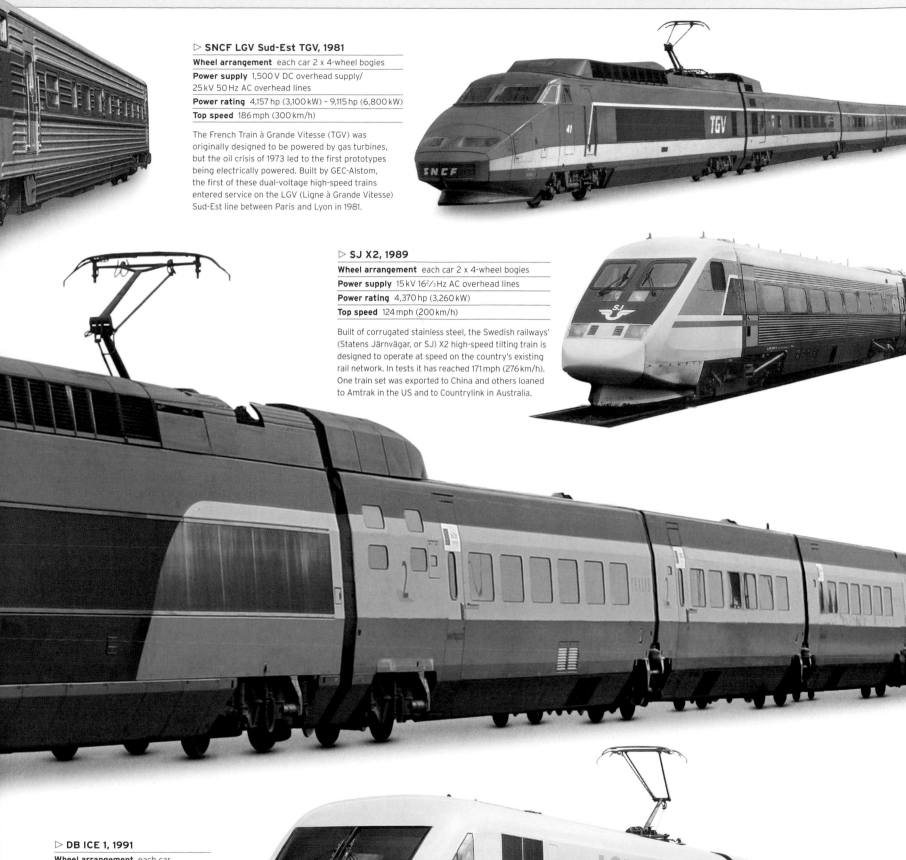

▷ **SNCF LGV Sud-Est TGV, 1981**

Wheel arrangement each car 2 x 4-wheel bogies

Power supply 1,500 V DC overhead supply/ 25 kV 50 Hz AC overhead lines

Power rating 4,157 hp (3,100 kW) – 9,115 hp (6,800 kW)

Top speed 186 mph (300 km/h)

The French Train à Grande Vitesse (TGV) was originally designed to be powered by gas turbines, but the oil crisis of 1973 led to the first prototypes being electrically powered. Built by GEC-Alstom, the first of these dual-voltage high-speed trains entered service on the LGV (Ligne à Grande Vitesse) Sud-Est line between Paris and Lyon in 1981.

▷ **SJ X2, 1989**

Wheel arrangement each car 2 x 4-wheel bogies

Power supply 15 kV 16²/₃ Hz AC overhead lines

Power rating 4,370 hp (3,260 kW)

Top speed 124 mph (200 km/h)

Built of corrugated stainless steel, the Swedish railways' (Statens Järnvägar, or SJ) X2 high-speed tilting train is designed to operate at speed on the country's existing rail network. In tests it has reached 171 mph (276 km/h). One train set was exported to China and others loaned to Amtrak in the US and to Countrylink in Australia.

▷ **DB ICE 1, 1991**

Wheel arrangement each car 2 x 4-wheel bogies

Power supply 15 kV 16³/₄ Hz AC, overhead supply

Power rating 5,094 hp (3,800 kW)– 6,437 hp (4,800 kW)

Top speed 174 mph (280 km/h)

Introduced in 1991, InterCityExpress (ICE) 1 was Germany's first truly high-speed public train. Sixty train sets were built, each one consisting of a power car at either end and either 12 or 14 passenger cars; a 12-car set can accommodate 743 passengers.

Building Great Railways
Eurostar

The modern era of high-speed rail travel between London, Brussels, and Paris began with the opening of the Channel Tunnel in 1994. However, on the English side of the channel, the new Eurostar trains were forced to run on a Victorian railway system until the full completion of the HS1 link in 2007.

London ●▨▨▨▨▨▨▨

London St Pancras ⬚1
After initially operating from London Waterloo from 1994, the Eurostar terminal was relocated to the refurbished London St Pancras International in 2007 with the opening of the HS1 link.

THE IDEA OF A TUNNEL under the English Channel to link the UK and France was not new. Various proposals were made during the 19th and early 20th centuries, but British fears that a tunnel could be used by an army invading England scuppered most plans, even though one 1929 design included a system for flooding the tunnel to repel invaders.

It was not until the 1960s that the French and British governments agreed to a modern project. Construction finally began in 1974, but halted within a year when the British, seeing costs soar and the economy crumble, cancelled the project. Eventually, in 1986, a private consortium of British and French banks and construction firms agreed to build the tunnel, work beginning from both sides in 1988. Two years later, the two ends of the service tunnel met under the channel.

Opened in 1994, the Channel Tunnel extends for 31⅓ miles (50.45 km) between Folkestone in England and Coquelles, near Calais in France. It consists of two single-track rail tunnels separated by a service tunnel, which can be used for passenger evacuation in an emergency. It is not a rail tunnel but a roadway where the tunnel maintenance crews use zero-emissions electric vehicles.

The new Eurostar service between Paris and London demanded high-speed railway lines and the French were first off the mark, opening the

1929 POSTER FOR PRE-EUROSTAR SERVICES

Eurostar at St Pancras International
St Pancras took over from Waterloo as London's international terminal in 2007, with the inauguration of the high-speed HS1 link from the Channel Tunnel.

207-mile (333-km) LGV Nord in 1993. This electrified line connects Gare du Nord, Paris to the Belgian border and the Channel Tunnel via Lille. The Belgians followed with their 55-mile (88-km) HSL 1 which opened in 1997, linking LGV Nord to Brussels-South.

In England, the Eurostar trains ran at lower speeds on existing railways between Folkestone in Kent and into special platforms at London Waterloo, a busy commuter station. Services commenced to Gare du Nord in Paris and Brussels-South station on 14 November 1994. It was to be a further 13 years before Britain's new 67-mile (108-km) High Speed 1 (HS1) line between Folkestone and the newly refurbished St Pancras International station in London opened on 14 November 2007. This reduced the journey time between London and Brussels to 1 hour 51 minutes. London to Paris took just 2 hours 15 minutes, more than four hours faster than when passengers had to disembark, cross the channel by ferry, and then board another train bound for the capital on the French side.

Speed restrictions
The series 373000 TGV (BR Class 373 in the UK) reaches high speeds in the countryside, but is restricted to 99 mph (160 km/h) in the Channel Tunnel.

KEY FACTS

DATES
1988 February: Channel Tunnel building and tunnelling begins
1990 December: the French and British tunnels meet underground
1993 June: the first Eurostar test train travels through the tunnel from France to the UK

TRAIN
Train set *Inter-Capital* (31 sets built, 27 in Eurostar service): 18 passenger carriages; 1,293 ft (394 m) long, capacity 750
Train set *North of London* or *Regional* (7 sets built, on long-term lease to SNCF): 14 passenger carriages; 1,050 ft (320 m) long, capacity 558
Train set *Nightstar* International service intended to run beyond London. Cancelled 1999 – all 139 coaches sold to Via Rail in Canada
Locomotives 27 Eurostar electric multiple unit (EMU) sets currently in service, Class 373/1 (UK) and TGV373000 (France). 2 power cars per set
Carriages 3 Eurostar travel classes: business premiere, standard premiere, and standard
Speed 186 mph (300 km/h) on high-speed lines; 99 mph (160 km/h) in the Channel Tunnel

JOURNEY
London St Pancras to Gare du Nord, Paris
305 miles (492 km); 2 hours 15 minutes (from 2007)
London St Pancras to Brussels-South 232 miles (373 km); 1 hour 51 minutes (from 2007)

RAILWAY
Gauge Standard gauge 4 ft 8 ½ in (1.435 m), cleared to larger European loading gauge
Channel Tunnel World's second-longest tunnel at 31 ⅓ miles (50.45 km) and longest undersea rail tunnel in the world at 23 ½ miles (37.9 km)
Bridges Medway Viaduct, UK, 4,265 ft (1.3 km)
Lowest point 250 ft (76 m) below sea level

2 **3** **Boring the tunnel**
The UK Crossover Tunnel was constructed almost 5 miles (8 km) from the UK coast, enabling trains to switch from one track to the other. At this point the entire South Rail boring machine could pass through on the way to France.

4 **Linking of England and France**
The undersea breakthrough of UK and French tunnels took place on 1 December 1990. The moment was later commemorated in the Channel Tunnel.

ENGLISH CHANNEL

Folkestone

BELGIUM

Calais

Brussels

Lille

Shuttle service Cars and trucks are transported between Folkestone and Coquelles, west of Calais, on *Le Shuttle* trains.

Brussels-South Station
Opened in 1952, the station has six platforms and is now only 1 hour 51 minutes from London.

French Crossover Cavern
A huge undersea cavern was also constructed 5 miles (8 km) off the French coast. These are the largest undersea caverns ever built.

Lille-Europe Station Opened in 1993 to serve Eurostar, TGV and other high-speed trains, Lille-Europe is just 54 minutes from Paris.

High-speed derailment Eurostar shares the LGV Nord with TGV services. In 1993 a TGV derailed at 183 mph (294 km/h) because of subsidence beneath the track, thought to be caused by World War I trench excavations. There were no serious injuries.

FRANCE

Through France Eurostar trains travel at 186 mph (300 km/h) on the LGV Nord, the first of the high-speed Channel Tunnel rail links to open in 1993.

Paris

5 **Gare du Nord, Paris**
Opened in 1846, Gare du Nord has been served by Eurostar since 1994. Four of the station's 44 platforms are devoted to Eurostar services.

KEY
● Start/Finish
● Main stations
▥▥▥ Main route
▥▥▥ Tunnel

| 0 | 25 | 50 miles |
| 0 | 25 | 50 | 75 km |

N

UK AND FRANCE UNITED

More than 13,000 engineers, technicians, and workers laboured to link the UK with France; and 11 tunnel-boring machines were used. Since 1994 use of the tunnel has grown until it now carries more passengers between London, Paris, and Brussels than all airlines combined.

Diesel's Next Generation

By the early 1980s the first home-grown generation of diesel–electric locomotives in Europe and North America had reached the end of their working lives. The US engine builders General Electric and the General Motors's EMD brand then began to dominate the scene on both continents with their highly successful, more powerful and efficient heavy-freight machines, which remain in operation today. In the UK, on the other hand, diesel locomotive building ended completely in 1987 when the last engine, BR Class 58 diesel–electric No. 58 050, rolled off the production line at the famous Doncaster Works.

△ **BR Class 58, 1984**

Wheel arrangement	Co-Co
Transmission	electric
Engine	Ruston Paxman 12-cylinder diesel
Total power output	3,300 hp (2,460 kW)
Top speed	80 mph (129 km/h)

Designed with an optimistic eye on export potential, 50 of the Class 58 heavy-freight, diesel-electric locomotives were built by British Rail Engineering Ltd at Doncaster between 1983 and 1987. They had a short working life in Britain with the last retired in 2002. Since then 30 have been hired for railways in the Netherlands, France, and Spain.

△ **Amtrak GE Genesis, 1992**

Wheel arrangement	B-B
Transmission	electric
Engine	General Electric V12 or V16 4-stroke supercharged diesel
Total power output	4,250 hp (3,170 kW)
Top speed	110 mph (177 km/h)

General Electric Transportation Systems built 321 of these low-profile, lightweight, diesel-electric locomotives between 1992 and 2001. They operate most of Amtrak's long-haul and high-speed rail services in the US and Canada. A dual-mode version can also collect 750 v DC current from third-rail in built-up areas such as New York.

△ **IÉ Class 201, 1994**

Wheel arrangement	Co-Co
Transmission	electric
Engine	EMD V12 2-stroke diesel
Total power output	3,200 hp (2,386 kW)
Top speed	102 mph (164 km/h)

Thirty-two of these powerful diesel-electric locomotives were built by General Motors in Ontario, Canada, for Iarnród Éireann in Ireland between 1994 and 1995. Two were also built for Northern Ireland Railways. They are all named after Irish rivers and operate on the Dublin to Cork express trains and on the *Enterprise* between Dublin and Belfast.

▷ **UP GM EMD Class SD60, 1984**

Wheel arrangement	C-C
Transmission	electric
Engine	EMD 16-cylinder diesel
Total power output	3,800 hp (2,834 kW)
Top speed	65 mph (105 km/h)

Built by General Motors, the heavy freight EMD Class SD60 diesel-electric locomotive was introduced in 1984. Production ceased in 1995 by which time 1,140 had been delivered to nine US railways, Canadian National Railways, and Brazil. Union Pacific Railroad bought 85 of the SD60, seen here, and 281 of the SD60M variant.

◁ **BR GM EMD Class 66, 1998**

Wheel arrangement Co-Co

Transmission electric

Engine EMD V12 two-stroke diesel

Total power output 3,000 hp (2,238 kW)

Top speed 75 mph (121 km/h)

A total of 446 of these diesel-electric freight locomotives were built by Electro-Motive Diesel in the US for Britain's railways between 1998 and 2008. Over 650 of this highly successful design have also been sold to several European freight operators as well as the Egyptian State Railways.

▷ **DWA Class 670 railcar, 1996**

Wheel arrangement 2-axle

Transmission mechanical

Engine MTU 6V 183 TD 13 diesel

Total power output 335 hp (250 kW)

Top speed 62 mph (100 km/h)

Incorporating parts used in buses, six of these double-deck diesel railcars were built by German Waqon AG (DWA) for German state railways in 1996 after a prototype was unveiled in 1994. A number remain in service.

◁ **ADtranz DE AC33C, 1996**

Wheel arrangement Co-Co

Transmission electric

Engine General Electric V12 diesel

Total power output 3,300 hp (2,462 kW)

Top speed 75 mph (121 km/h)

Fitted with General Electric diesel engines these powerful locomotives, nicknamed "Blue Tigers", were built by German manufacturer ADtranz between 1996 and 2004. Eleven units, including No. 250 001-5 seen here, were made for leasing in Germany, while Pakistan Railways ordered 30 and Keretapi Tanah Melayu in Malaysia bought 20.

▷ **HSB Halberstadt railcar, 1998**

Wheel arrangement 2 x 4-wheel bogies (1 powered)

Transmission mechanical

Engine Cummins 6-cylinder 1,080 cc diesel

Total power output approx 375 hp (280 kW)

Top speed 31 mph (50 km/h)

Four of these were built in 1999 by the Halberstadt Works, then part of Deutsche Bahn, for the Harzer Schmalspurbahnen (Harz Narrow-gauge Railway). They still work services at times, running on lines that are lightly used.

A New Wave of Electrics

The demise of steam power in Western Europe during the 1950s and 1960s saw the spread of electrification across much of the continent. The soaring price of oil in the 1970s added further impetus for national railways to switch from hurriedly introduced diesel locomotives to electric haulage. However, the power supplies varied greatly from country to country, and with the growth of transnational railway freight services, a new generation of multivoltage electric locomotives had started to appear by the 1990s.

▷ CSD Class 363, 1980

Wheel arrangement	B-B
Power supply	25 kV 50 Hz AC/3,000 V DC, overhead supplies
Power rating	4,102–4,666 hp (3,060–3,480 kW)
Top speed	75 mph (121 km/h)

The prototype Class 363 dual-voltage locomotive was built by Skoda Works for the Czechoslovakian state railways. It was the first multisystem electric engine in the world fitted with power thyristor pulse regulation and has a distinct sound in three frequencies when accelerating.

◁ DR Class 243, 1982

Wheel arrangement	Bo-Bo
Power supply	15 kV 16.7 Hz AC, overhead supply
Power rating	4,958 hp (3,721 kW)
Top speed	75 mph (120 km/h)

Over 600 of these mixed-traffic electric locomotives were built by L.E.W. Hennigsdorf for the Deutsche Reichsbahn between 1982 and 1991. Originally classified as DR Class 243, they became Class 143 under the renumbering scheme that followed Germany's reunification.

△ PKP Class EP 09, 1986

Wheel arrangement	Bo-Bo
Power supply	3,000 V DC, overhead supply
Power rating	3,914 hp (2,920 kW)
Top speed	99 mph (160 km/h)

A total of 47 of the Class EP09 express passenger electric engines were built by Pafawag of Wroclaw for the Polish state railways between 1986 and 1997. First entering service in 1988, they operate trains on main lines from Warsaw and Kraków.

△ SNCF Class BB 26000, 1988

Wheel arrangement	B-B
Power supply	25 kV AC/1,500 V DC, overhead supplies
Power rating	7,500 hp (5,595 kW)
Top speed	124 mph (200 km/h)

These multipurpose, dual-voltage electric engines were constructed for the French state railways between 1988 and 1998; a total of 234 were built. A further 60 triple-voltage locomotives, which were made between 1996 and 2001, are classified as SNCF Class BB 36000.

TECHNOLOGY

Glacier Express

Named in honour of the Rhone Glacier, which it passed at the Furka Pass, the Glacier Express was introduced between St Moritz and Zermatt in Switzerland on 25 June 1930. It was originally operated by three 3-ft 3-in- (1-m-) gauge railway companies, the Brig-Visp-Zermatt Bahn (BVZ), the Furka Oberalp Bahn (FO), and the Rhaetian Railway (RhB). While two of the lines were electrified, steam locomotives were used on the FO section until 1942 when that line was also electrified. It runs daily all-year-round but is not exactly an "express" as it takes 7½ hours to cover 181 miles (291 km), much of it on a rack-and-pinion system. Since 2008 much of its route on the Albula and Bernina railways has been declared a UNESCO World Heritage Site.

Scenic ride The train passes through stunning Alpine scenery, crossing 291 bridges, burrowing through 91 tunnels, and gaining height on numerous spirals.

△ BR Class 91, 1988

Wheel arrangement Bo-Bo

Power supply 25 kV AC, overhead supply

Power rating 6,480 hp (4,832 kW)

Top speed 125 mph (204 km/h)

Delivered between 1988 and 1991, 31 of the Class 91 express locomotives were built at Crewe Works for British Rail. Designed to reach 140 mph (225 km/h) but now only used at 125 mph (204 km/h), they operate express trains in a push-pull mode on the East Coast Main Line between London King's Cross and Edinburgh.

△ FS Class ETR 500, 1992

Wheel arrangement power cars: 2 x 4-wheel motorized bogies

Power supply 3 kV DC, overhead supply

Power rating complete train: 11,796 hp (8,800 kW)

Top speed 155 mph (250 km/h)

Following four years of testing, 30 Class ETR 500 high-speed, single-voltage electric trains were introduced on the Italian state railway, between 1992 and 1996. Before the production models were constructed, a prototype motor car was built and tested. Coupled to an E444 locomotive on the Diretissima Line between Florence and Rome, it attained a speed of 198 mph (319 km/h) in 1988.

◁ SBB Cargo Bombardier Traxx, 1996

Wheel arrangement Bo-Bo

Power supply 15 kV 16.7 Hz AC/25 kV 50 Hz AC, overhead supply

Power rating 7,500 hp (5,595 kW)

Top speed 87 mph (140 km/h)

From 1996 the Bombardier Traxx, dual-voltage electric locomotives were introduced on many European railways. Since then around 1,000 have been built at the company's assembly plant in Kassel, Germany, of which 35 of the F140 AC variant, seen here, are operated by SBB Cargo in Switzerland.

▽ BR Class 92, 1993

Wheel arrangement Co-Co

Power supply 25 kV AC, overhead supply/750 V DC third-rail

Power rating 5,360-6,760 hp (3,998-5,041 kW)

Top speed 87 mph (140 km/h)

Designed to haul freight trains through the Channel Tunnel between Britain and France, the 46 Class 92, dual-voltage electric locomotives were built by Brush Traction and ABB Traction and assembled at the former company's erecting shops in Loughborough (UK) between 1993 and 1996. They are operated by GB Railfreight/Europorte 2 and DB Schenker.

▷ Amtrak Class HHP-8, 1999

Wheel arrangement B-B

Power supply 12.5 kV 25 Hz AC/12.5 kV 60 Hz AC/25 kV 60 Hz AC, overhead supplies

Power rating 8,000 hp (5,968 kW)

Top speed 125 mph (201 km/h)

Fifteen of these express passenger electric locomotives were built for Amtrak by Bombardier and Alstom in 1999. The Amtrak locomotives hauled trains on the Northeast Corridor between Washington DC and Boston until they were retired in 2012.

Palace on Wheels

Travelling in the style of a Maharaja through India's most evocative destinations is one of life's most luxurious railway experiences. The *Palace on Wheels*, one of the world's top five luxury trains, is a reconstruction based on the stately personal carriages of the rulers of Rajasthan and Gujarat, the Nizams of Hyderabad, and the Viceroys of India. The original carriages were in use from 1917 until India left the British Empire in 1947.

THE ORIGINAL HIGHLY ORNATE "royal" carriages, furnished in antique silk, were deemed inappropriate for India's fleet of standard passenger trains and put out of commission. However, in 1982, Indian Railways teamed up with the Rajasthan Tourist Development Cooperation to provide a new luxury metre-gauge service with an ivory livery and plush carriages that emulated the grand decor of an earlier age. Powered by a steam engine, the new Palace on Wheels made its maiden journey on January 26, the anniversary of India's republic.

In the 1990s there was a further re-invention of the train when the railway switched to broad gauge. The Palace on Wheels accommodation was replaced with modern air-conditioned cabins with attached bathrooms, each saloon named after one of the royal provinces of Rajasthan, with interiors that reflected the history of the region through paintings, furniture and handicrafts. Still in use today, the train is made up of 14 saloons, a kitchen car, two restaurants, a bar with a lounge, and four service cars. To add a further touch of majesty to the experience, the train offers personal "Khidmatgars", or attendants, who are available to serve guests around the clock.

The seven-day round trip on the Palace on Wheels has since become a major tourist attraction that draws people from around the world. The journey for 80 passengers begins in New Delhi and travels through major sites in northwest India's golden triangle, taking in wildlife safari parks and ending at the Taj Mahal.

STEAM ENGINE

DIESEL ENGINE

Aspiring to royalty
The *Palace on Wheels* train is a spectacular re-creation of the royal and official trains of the Indian 3-ft 3-in- (1-m-) gauge in the 1920s and 1930s. Today, its carriages are rarely hauled by steam - most journeys are powered by a diesel locomotive.

SPECIFICATIONS FOR CARRIAGES	
Origin	India
In-service	1982–present
Coaches	14
Passenger capacity	approx. 80
Route	Rajasthan and the Golden Triangle (Delhi–Jaipur–Agra)

Trip of a lifetime
The week-long trip on the *Palace on Wheels* takes passengers through northwestern India on a nostalgic journey to some of the most popular tourist spots in the Golden Triangle.

Windows run the full length of the carriage

Exterior paintwork is identical on every saloon carriage

Coat of arms identifies the princely state that inspired the interior decoration

Painted sign shows name of saloon car

Passenger doors at each end with stylized oval windows

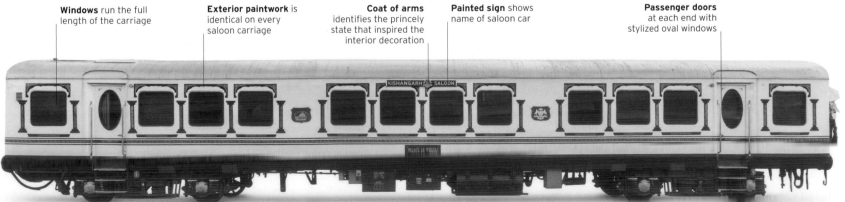

KISHANGARH SALOON CARRIAGE

Fit for a king
The two restaurants on board the *Palace on Wheels* are called *Maharaja* (shown here) and *Maharani*. An original royal dining saloon is kept in the National Railway Museum in Delhi.

JAIPUR SALOON AND BEDROOM

The Jaipur saloon is decorated in colours that represent the former Rajput state of Jaipur, while the exterior of the carriage bears its coat of arms. The ceiling is adorned in the region's famed "phad" (foil work) and illustrates religious festivals such as Teej, Holi, Gangaur, and Diwali. Each saloon consists of four coupes (sleeping rooms) and a bathroom. A mini pantry and a lounge provide additional comfort.

1. Name of carriage embossed on metal plate **2.** "Phad" (foil work) on ceiling depicting festivals celebrated in Rajasthan **3.** Glass and gilt ceiling light **4.** Saloon with banquet-style sofas and painted fresco ceiling **5.** Metal hand plate on door **6.** Carriage corridor **7.** Coupe (sleeping room) **8.** Mirror inside the coupe **9.** Switches for lights and music **10.** Ensuite facilities with elegant modern fittings and mirror

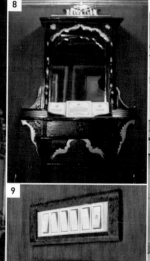

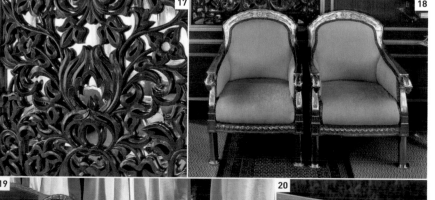

PALACE ON WHEELS BAR

The lounge bar is designed to reflect a contemporary royal style with flourishes that hark back to the Rajput era. Made of wood, marble, and brass fixtures, the bar area epitomizes the aesthetic of the time. A selection of antique pitcher designs ornament the front of the counter area, depicting some of the drink-pouring vessels the maharajas would have used.

11. Bar and lounge carriage **12.** Marble-top bar counter **13.** Antique pitcher design in marble, with gold inlay work, on front of bar counter **14.** Emergency stop chain **15.** Chandelier **16.** Peacock motif in tinted glass **17.** "Jaali" (teak latticework) panel **18.** Armchairs with "patra" (oxidized white metal) work on borders **19.** Deep-cushioned sofa with raw silk upholstery **20.** Intricately carved elephant head design – a sign of prosperity – at end of armrest

MAHARANI RESTAURANT

The Rajasthani theme continues in the interior design of the Maharani (meaning "Queen") dining cabin, with floral carpets and curtains, and featuring framed art from the Mughal period hanging on the walls. The most opulent touch is arguably the mirrored and teakwood-panelled ceiling.

1. Sumptuous dining room with mirrored ceiling **2.** Mughal art in marble, created with vegetable colours, on carriage wall **3.** Silk-embroidered drapes with floral design **4.** Tree motif in stained glass on restaurant door **5.** Panelled corridor **6.** Kitchen positioned at one end of restaurant carriage

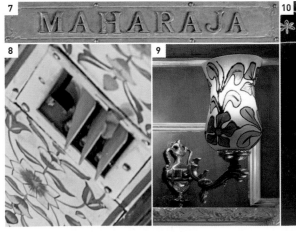

MAHARAJA RESTAURANT

Maharaja means "King" in Hindi, and accordingly, this dining carriage has a more masculine feel compared to the Maharani. Drapes of royal blue adorn the elegant, mahogany-led decor. The seating is arranged in groups of four. Both restaurants serve different varieties of cuisine, although there is an emphasis on Rajasthani dishes.

7. Name plaque above the door **8.** Air vent in central ceiling panel **9.** Wall light with painted glass shade **10.** Gold-embroidered *zari* work on velvet drape **11.** Restaurant carriage decorated with mahogany panelling

ROYAL SPA

The *Palace on Wheels* has brought its luxury service up to date with the recent addition of a carriage dedicated to spa services, equipped with state-of-the art equipment in a relaxing modern setting. Passengers can enjoy massage or a range of revitalizing treatments as they speed through the countryside to their next heritage destination.

12. Corridor in the Royal Spa **13.** Double-bed massage suite **14.** Tip-back seat and basin **15.** Pedicure bowl with rose petals

GENERATOR CAR AND GUARD'S COMPARTMENT

The guard's compartment and generator car are located at the front of the train, away from the palatial setting of the passenger carriages. The generator provides the electricity necessary to power the lights, appliances, kitchen, and bar equipment. In the guard's cabin, a close eye is kept on gauge and meter readings to ensure the train runs smoothly and that passengers have a comfortable journey.

16. Power control panel in the generator car **17.** Guard's compartment
18. Handbrake **19.** Temperature control panel **20.** Vent control **21.** Air brake

Urban Rail Solutions

While pioneering urban railways such as London's Metropolitan Railway and Chicago's South Side Elevated Railroad originally ran on steam, by the late 1930s electrified railways such as the Budapest Metro, the Moscow Metro and London Underground were carrying huge numbers of commuters between their suburban homes and city-centre offices. With the world's cities still expanding during the late 20th century, modern electrically powered rapid-transit systems (RTS) such as street tramways and surface and underground railways, many using driverless automatic trains, were built to transport millions of passengers each day, very quickly and over short distances.

▷ **Vancouver SkyTrain RTS ICTS Mark I, 1985**

Wheel arrangement	2-car sets
Power supply	750 V DC, third rail
Power rating	888 hp (640 kW) per 2-car set
Top speed	50 mph (80 km/h)

The 43-mile (69-km) Vancouver SkyTrain is an RTS serving Vancouver and its suburbs. Trains on the Expo Line and Millennium Line are automated and are driven by linear induction motors. The cars run in two- to six-car configurations. This is a four-car Mark I train of the Intermediate Capacity Transit System (ICTS) built by the Urban Transportation Development Corporation of Ontario.

△ **SDTI Duewag U2 cars, 1980/81**

Wheel arrangement	double-ended, 6-axle, articulated
Power supply	600 V DC, overhead supply
Power rating	408 hp (300 kW)
Top speed	50 mph (80 km/h)

The San Diego Trolley is a 53-station, three-route light rail system in the city of San Diego, California, which opened in 1981. The articulated cars initially used were U2 vehicles built in Germany by Duewag. These cars also worked in Edmonton and Calgary, Canada, and in Frankfurt, Germany.

△ **Berlin *U-Bahn* F-type train, 1992/1993**

Wheel arrangement	2-car sets
Power supply	750 V DC, third rail
Power rating	734 hp (540 kW) per 2-car set
Top speed	45 mph (72 km/h)

The 152-mile (245-km) Berlin *U-Bahn* (or underground railway), first opened in 1902 and, despite problems caused by the division of Berlin during the Cold War, today serves 170 stations across 10 lines. The system uses both *Kleinprofil* (small profile) trains and *Grossprofil* (large profile) trains, such as the F-type. The trains are worked in four-, six-, or eight-car combinations.

△ T&W Metro, 1980

Wheel arrangement 2-car articulated (6-axle articulated sets)

Power supply 1,500 V DC, overhead supply

Power rating 410 hp (301.5 kW)

Top speed 50 mph (80 km/h)

The 46-mile (74-km) Tyne & Wear Metro in Newcastle-upon-Tyne in northeast England is a hybrid light railway system with suburban, interurban, and underground sections. A total of 90 two-car articulated sets, usually coupled together in pairs, were built between 1978 and 1981 by Metro Cammell in Birmingham.

▽ Vienna ULF tram, 1998

Wheel arrangement 2- or 3-car articulated

Power supply 216–480 kW (289–643 hp)

Power rating 653 hp (480 kW)

Top speed 50 mph (80 km/h)

The Ultra Low Floor (ULF) cars, built by the consortium of Siemens of Germany and Elin of Austria, were introduced on Vienna's tram network in 1998 and in 2008 in Oradea, Romania. With a floor only 7 in (18 cm) above the pavement they provide easy access for wheelchairs or children's buggies.

▽ Luas Alstom Citadis tram, 1997

Wheel arrangement 3-, 5-, and 7-car articulated (8-, 12-, and 16-axle articulated sets)

Power supply 750 V DC, overhead lines

Power rating 979 hp (720 kW)

Top speed 44 mph (70 km/h)

The Citadis is a family of low-floor trams built by Alstom in France and Spain and popular in many cities around the world. The 23-mile (37-km) Luas tram system in Dublin, Ireland, uses the three-car 301 and five-car 401 variants on the city's Red Line, while the seven-car 402 variant works on the Green Line.

△ SMRT North-South Line C151, 1987

Wheel arrangement 6-car sets

Power supply 750 V DC, third rail/ 1,500 V DC, overhead supply

Power rating 2,937 hp (2,160 kW)

Top speed 50 mph (80 km/h)

Singapore's Mass Rapid Transit system started life when the North–South Line opened in 1987. Since then it has been extended to 93 miles (150 km), serving 106 stations on five routes. Six-car C151 (shown), C151A, and C751B trains collect current from a third rail and have an automatic train operation system. C751A trains, which are fully automatic and driverless, use an overhead supply.

▽ Gatwick Adtranz C-100, 1987

Wheel arrangement 2-car sets with rubber tyres

Power supply 600 V AC

Power rating 110.5 hp (75 kW) per car

Top speed 28 mph (46 km/h)

This elevated, fully automatic, driverless, guided people-mover system began operation in 1987. The train connects the North and South Terminals at London's Gatwick Airport, a distance of $^3/_4$ mile (1.2 km). Similar systems have proved popular in airports and cities around the world.

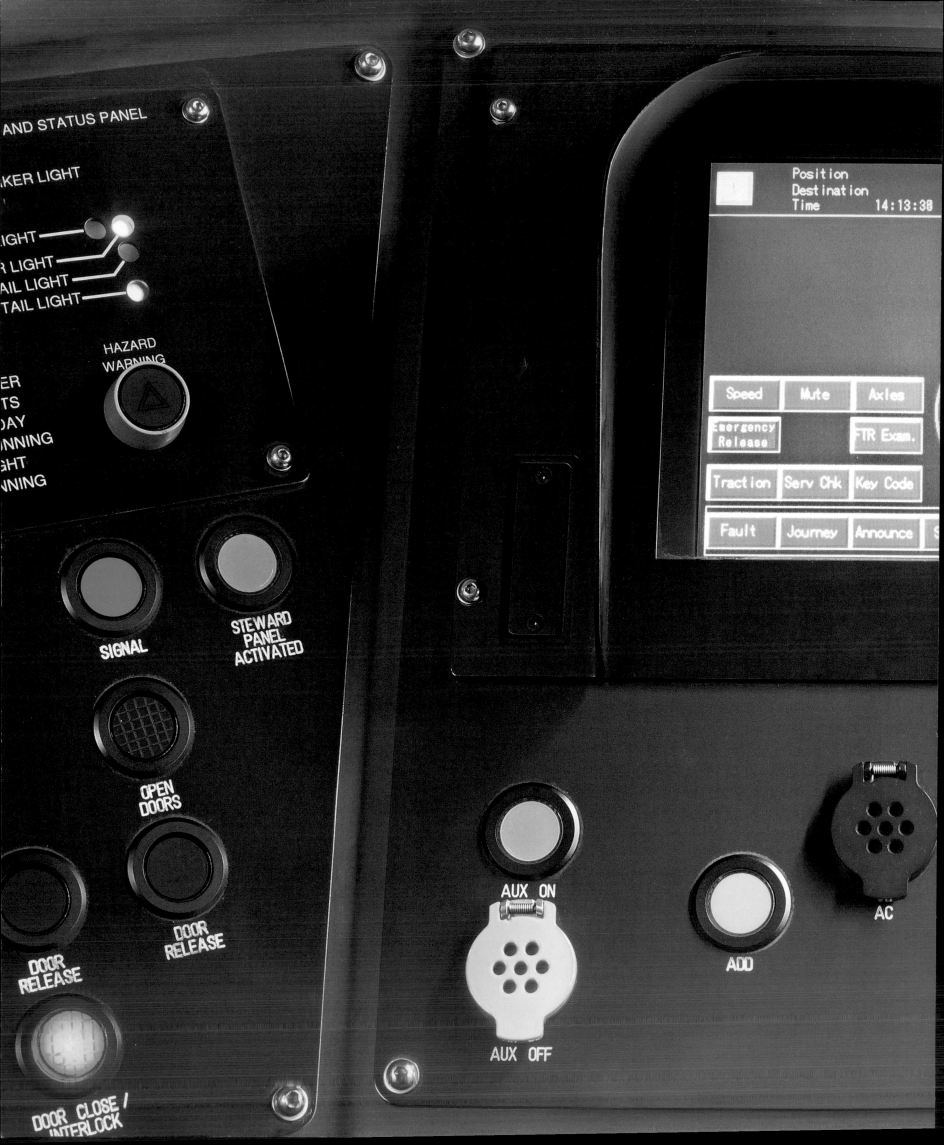

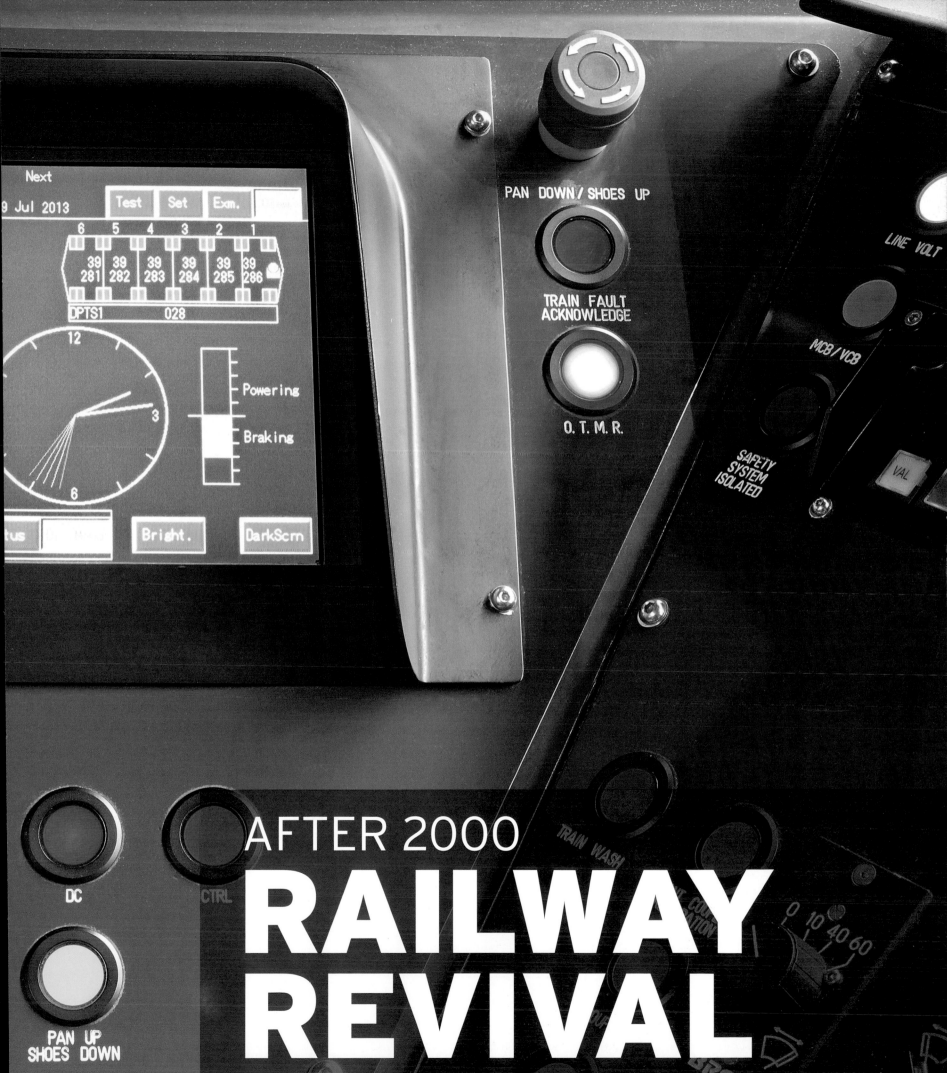

AFTER 2000

RAILWAY REVIVAL

RAILWAY REVIVAL

As the world looks for alternatives to roads, rail has once again become a priority. New energy-efficient locomotives have been launched and the search for even higher speeds continues. In 2003 an experimental Maglev (*magnetic levitation*) train in Japan reached a speed of 361 mph (581 km/h) - the world's first commercial high-speed Maglev began operations in Shanghai, China, in 2004. Two years later, the opening of the railway to Lhasa in Tibet broke a different kind of record, taking normal trains to altitudes never before reached. Oxygen masks are available to passengers on trains that operate at more than 16,000 ft (5,000 m). In 2007 a specially adapted TGV in France broke the world speed record for a conventional train.

Expansion of high-speed rail has continued around the world. China has opened thousands of miles of track to create the world's largest high-speed network. Spain, which entered the new era in 1992, has set out to create Europe's biggest system. The launch of the *Acela Express* in the US pushed the nation's maximum speed up to 150 mph (241 km/h), while the UK completed a dedicated high-speed line connecting London and the Channel Tunnel.

Yet progress is not all about going faster. Metros and light rail have continued to expand their reach - and operators have pressed ahead with greater automation. As the 21st century's first decade drew to a close, the Dubai Metro became the world's longest fully automated line, at 47 miles (75 km).

Rail travel has reinvented itself as a luxurious alternative to cramped aircraft or gridlocked roads - passengers can travel through ever-changing landscapes in style. More than two centuries after it began, the railway era is far from over.

> " Any railway, **working properly,** is a marvel of **civilized** co-operation"

LIBBY PURVES, *THE TIMES*, 14 MAY, 2002

THE VENICE SIMPLON-ORIENT-EXPRESS
...A JOURNEY LIKE NO OTHER

vsoe.com

△ **Orient Express luxury travel**
This modern poster for the Venice Simplon-Orient Express features a liveried porter and hints at a return to the luxury travel of a bygone era.

◁ **A "Bullet Train"** speeds through a district of high-rise office blocks in Tokyo, Japan

Key Events

▷ **2000** The *Acela Express* is introduced in the US. It reaches speeds of up to 150 mph (241 km/h).

▷ **2003** The first section of the UK's high-speed Channel Tunnel Rail Link is opened. It reaches London in 2007 and is now known as High Speed 1 (HS1).

▷ **2003** An experimental Maglev train reaches 361 mph (581 km/h) in Japan, a new world record for a manned train.

▷ **2004** The world's first commercial high-speed Maglev system opens in Shanghai, China.

△ **Shanghai Transrapid Maglev train**
The world's only Maglev train service runs between Shanghai and the city's Pudong International Airport. It reaches speeds of 268 mph (431 km/h).

▷ **2006** Services start on the world's highest conventional railway. The route in Tibet reaches up to 16,640 ft (5,072 m) above sea level.

▷ **2007** An experimental French TGV sets a new world record for conventional-wheeled trains of 357¼ mph (575 km/h).

▷ **2007** China enters the modern high-speed rail age with a new dedicated line.

▷ **2009** The first section of Dubai's metro opens, followed by another in 2010; at 47 miles (75 km) it is the world's longest fully automated metro.

▷ **2012** Completion of the major sections of the Beijing to Hong Kong high-speed line makes it the world's longest. By the time it is finished in 2015 it will be around 1,450 miles (2,234 km) long.

Universal Applications

The beginning of the 21st century saw changes in the way manufacturers dealt with their customers – instead of railway companies telling the equipment manufacturer exactly what they wanted built, the manufacturers started offering railway operators product ranges based on universal "platforms" much like the auto or aviation industries. As a result some commuters in California now travel in similar trains to those in Berlin or Athens, and interoperable locomotives, able to run from multiple traction voltages and using several different signalling systems, are now common in Europe.

△ **Siemens Eurosprinter ES64 U2/U4, 2000**

Wheel arrangement	Bo-Bo
Power supply	1,500 V DC/3,000 V DC and 15 kV AC/25 kV AC, overhead lines
Power rating	8,579 hp (6,400 kW)
Top speed	143 mph (230 km/h)

Siemens introduced the Eurosprinter family of locomotives following big orders from DB in Germany and ÖBB in Austria. The Eurosprinter range has four basic bodyshells and multiple versions. The ES64 U4 (EuroSprinter 6400 kW Universal 4 system) is the most flexible and able to operate in multiple countries.

△ **Siemens Desiro Classic, 2000**

Wheel arrangement	2-car DMU
Transmission	mechanical
Engine	2 x MTU 1800 6R
Total power output	845 hp (630 kW)
Top speed	74 mph (120 km/h)

Siemens has now sold over 600 of its first "Desiro" model, the two-car articulated diesel-powered Desiro Classic. The trains are used for regional passenger services in Europe, and the design has been exported to southern California. Electric versions have also been built for Bulgaria, Greece, and Slovenia.

▽ **Voith Gravita, 2008**

Wheel arrangement	B-B
Transmission	hydraulic
Engine	8 V 4000 R43
Total power output	1,341 hp (1,000 kW)
Top speed	62 mph (100 km/h)

The Gravita family of locomotives, developed by Voith, is designed for freight traffic on lightly used lines. Germany's Deutsche Bahn purchased 130 locomotives of two types – 99 of the Gravita 10BB and 31 of the more powerful 15BB model.

▷ Siemens Desiro-RUS, 2013

Wheel arrangement 2-car EMU

Power supply 3,000 V DC and 25 kV AC, overhead lines

Power rating 3,418 hp (2,550 kW)

Top speed 99 mph (160 km/h)

Desiro EMUs have been built for several countries from the UK to Slovenia, Greece, and Thailand. Russian railways (RZD) ordered 38 Desiro-RUS to operate services at the 2014 Sochi Winter Olympics. The trains, branded "Lastochka" (swallow), were built by Siemens at their Krefeld factory in Germany.

◁ Bombardier ALP45 DP, 2012

Wheel arrangement Bo-Bo

Transmission electric

Engine 2 x Caterpillar 3512C

Power supply 25 kV and 12.5 kV AC, overhead wires

Total power output diesel: 4,200 hp (3,135 kW)/ electric: 5,362 hp (4,000 kW)

Top speed diesel: 100 mph (161 km/h)/ electric: 124 mph (200 km/h)

These engines were designed for through operation from busy electric lines to non-electrified regional routes in North America to facilitate "one-seat rides" – travelling without changing trains. The locomotive can switch from electric to diesel (and vice versa) while moving. Bombardier has sold 46 to New Jersey Transit and Agence Métropolitaine de Transport in Montreal, Canada.

▷ Voith Maxima, 2008

Wheel arrangement C-C

Transmission hydraulic

Engine ABC 16 V DZC

Total power output 4,826 hp (3,600 kW)

Top speed 75 mph (120 km/h)

Voith introduced its most powerful single-engine, diesel-hydraulic locomotive ever built for freight operators in Europe in 2008. Two versions are available – the Maxima 40CC and lower-powered Maxima 30CC. Around 20 have been sold, mostly to Germany-based operators.

△ Vossloh Eurolight, 2010

Wheel arrangement Bo-Bo

Transmission electric

Engine Caterpillar C175-16

Total power output 3,753 hp (2,800 kW)

Top speed 124 mph (200 km/h)

The Eurolight design from Vossloh aims to maximize available power while minimizing axle weight, which enables the locomotive to operate even on rural routes often not built for heavy trains. By using a lighter weight engine and lightweight body, the locomotive weighs under 78 tons (79 tonnes).

△ Vossloh G6, 2010

Wheel arrangement C

Transmission hydraulic

Engine Cummins QSK-23-L

Total power output 900 hp (671 kW)

Top speed 50 mph (80 km/h)

Vossloh builds the G6 diesel-hydraulic locomotive in Kiel in Germany at the former Maschinenbau Kiel (MaK) factory. So far it has been sold mostly to industrial operators in Germany. Verkehrsbetriebe Peine-Salzgitter (VPS) runs a large railway network serving the steel industry at Salzgitter (central Germany) and has bought 40 G6s to replace 43 older diesel shunters.

△ Alstom Prima II, 2010

Wheel arrangement Bo-Bo

Power supply 3,000 V DC and 25 kV AC, overhead lines

Power rating 5,630 hp (4,200 kW)

Top speed 124 mph (200 km/h)

Alstom developed the Prima II prototype in 2008 and has sold 20 to Moroccan railways (ONCF). The locomotives are used for passenger trains on all electrified routes. Able to work from all traction voltages, the Prima II model can be built as four-axle locomotives or six-axle freight versions.

Historic Railways

Although many now operate solely for the benefit of the tourist trade, today's historic and heritage railways were first created to fulfil specific industrial or commercial functions. While a few continue to perform these duties, for many, the original reason for building the railway has gone. Nowadays it often falls to railway enthusiasts to restore and preserve some of the world's most enchanting lines for posterity.

1. **The Durango & Silverton Narrow Gauge Railroad** (1881) in Colorado serviced gold and silver mines but is now a National Historic Landmark, running its original steam engines.   2. **The White Pass & Yukon Route** (1898) was Alaska's "railway built of gold" during the Klondike Gold Rush. Closed in 1982, it is now run as a tourist attraction.
3. **The Ferrocarril Chihuahua-Pacífico, "El Chepe",** (1961) traverses the Copper Canyon in Mexico. First planned in 1880, finances and the rugged landscape delayed construction.   4. **The Old Patagonian Express, "La Trochita"** (1935) in Argentina faced closure in 1992 but now runs more than 20 steam locomotives.   5. **The Furka Cogwheel Steam Railway** (1925) in Switzerland was abandoned when a mountain tunnel was built in 1982. Volunteers now operate trains up to Furka Station at 7,087 ft (2,160 m).
6. **Chemin de Fer de la Baie de la Somme** (1887) runs around part of the coast of northern France using vintage stock.   7. **The Historical Logging Switchback Railway** (1926) in Vychylovka, Slovenia, closed in 1971 but part of the line has been in operation for tourists since 1994.   8. **The Giant's Causeway & Bushmills Railway** (1883) in Ireland was a tramway powered by hydroelectricity. Closed in 1949, it reopened in 2002 using steam and diesel power.   9. **The Bluebell Railway** (1882) in the UK, formerly the Lewes and East Grinstead line, closed in 1958, reopening after two years as the world's first preserved standard-gauge passenger line.   10. **The Ventspils Narrow Gauge Railway** (1916) in Latvia was built by the German Army during World War I. Trains run today on a 1¼-mile (2-km) track.   11. **The Brocken Railway** (1898), or Brockenbahn, in Germany's Harz Mountains has run tourist steam trains to Brocken mountain peak at 3,743 ft (1,141 m) since 1992.
12. **The Puffing Billy Railway** (1900) once served farming and forestry near Melbourne, Australia, and is now preserved for tourists.s   13. **The Darjeeling Himalayan Railway** (1881) is listed by UNESCO as one of the most outstanding examples of a hill railway in the world.

8

11

12

9 13

10

Clan Line & Belmond British Pullman

Built in 1948, Merchant Navy Class No. 35028 *Clan Line* operated as a mainline express passenger locomotive until 1967. Since *Clan Line* returned to service in 1974 it has been running special trips on Britain's main lines, and regularly hauls the *Belmond British Pullman* train. It has had three major overhauls in that time.

DESIGNED BY OLIVER BULLEID, *Clan Line* is one of 30 Merchant Navy Class 4-6-2 Pacific locomotives built from 1941 at the Southern Railway's Eastleigh Works. Each one was named after shipping companies that worked at the railway's Southampton Docks. *Clan Line* worked on the Southern Region of the newly nationalized British Railways and in 1959 was rebuilt into its current form. In July 1967 it was sold to the Merchant Navy Locomotive Preservation Society, which later teamed *Clan Line* up with the *Belmond British Pullman* train.

The cars of the train once formed part of some of Britain's most famous services such as the *Brighton Belle* and the *Queen of Scots*. After being withdrawn from service in the 1960s and 1970s, many fell into disrepair. In 1977 James B. Sherwood began buying and restoring the historic sleepers, saloons, and restaurant cars with the aim of reviving the legendary Orient Express; he acquired 35 cars in all.

FRONT VIEW

REAR VIEW

Coal space was extended in the 1995 overhaul to increase coal capacity

Tender has a sloping edge to the top to allow the crew better vision when moving backwards

Cab designed by Bulleid in consultation with crews

Nameplate mounted on side

BR-type smoke deflectors fitted after the 1959 rebuild

Elegant locomotive
Clan Line is maintained in working order so that it can run on Britain's main lines. The entire class was rebuilt in the 1950s, so none survive in their as-built, "air-smoothed" condition.

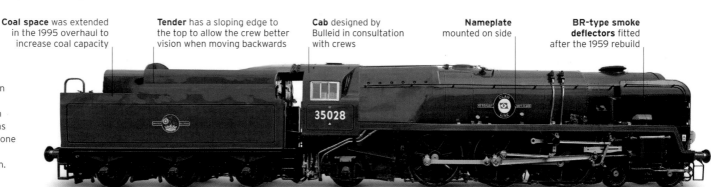

SPECIFICATIONS FOR CLAN LINE				
Class	Merchant Navy	**In-service period**	1948 to present (*Clan Line*)	
Wheel arrangement	4-6-2	**Cylinders**	3	
Origin	UK	**Boiler pressure**	280 psi (197 kg/sq cm) as built	
Designer/builder	Bulleid/Eastleigh Works	**Driving wheel diameter**	74 in (1,880 mm)	
Number produced	30 Merchant Navy Class	**Top speed**	105 mph (167 km/h)	

SPECIFICATION FOR CARRIAGES	
Origin	UK
In-service	various
Coaches	11
Passenger capacity	20-26 seats
Route	various

Hand rails on either side of the doors

Each car has two sets of 4-wheel bogies

Carriage names are displayed on the side

Pullman coat of arms displayed on the sides of each car

Palaces on wheels
The *Belmond British Pullman's* carriages run on the English leg of the famous Orient-Express transcontinental train. They were restored to meet rigorous safety standards, yet maintain their stunning vintage features.

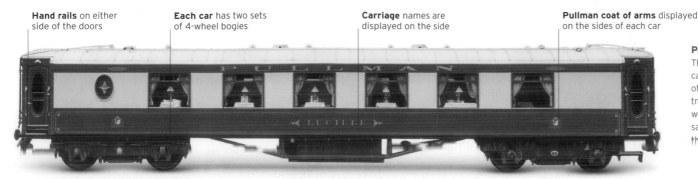

Travelling in style
Clan Line began its life hauling prestigious heavy boat trains with names such as *Golden Arrow* and *Night Ferry*. Nowadays *Clan Line* pulls the elegant carriages of the *Belmond British Pullman* train, which offers a movable feast of fine dining and silver service.

The later BR logo
Known as the "ferret and dartboard crest" British Railways used this logo on steam locomotives from 1956.

LOCOMOTIVE EXTERIOR

The Merchant Navy Class locomotives were originally built with straight-sided, cylinder cladding and an "air-smoothed", streamlined casing. The view from the cab was poor, and at speed the flat top generated a vacuum that drew exhaust steam down to obscure the driver's view. The engines were nicknamed "spam cans" because the shape of their casing resembled the tins of meat imported from the US at the time. The locomotives underwent a rebuild in the mid-50s emerging with a more conventional look, and the casing was replaced by conventional boiler cladding. *Clan Line* entered service painted in Southern Railway's Malachite Green livery, with British Railways' lettering.

1. Locomotive cab side number and power classification **2.** Headboard (when hauling the *Belmond British Pullman*) **3.** Electric headlight **4.** Draw hook and screw coupling **5.** Front buffer **6.** Right-hand compression valve and cylinder drain cock pipes **7.** Leading wheel **8.** Walschaerts valve gear – slide bar, crosshead, piston, and radius arm **9.** Locomotive lubricators **10.** Whistle **11.** Nameplate **12.** Rear and middle driving wheels **13.** Boiler water injector (live steam monitor type) **14.** Water injector delivery pipes to clack valves **15.** Tender axle-box **16.** Driving wheel brake block

CAB INTERIOR

The Merchant Navy Class was reputed to have one of the safest cabs. The fire door contained air holes that minimized the risk of blowbacks, while levers allowed the driver and fireman to operate the blowers without needing to step in front of the firehole.

17. General arrangement of driver's cab **18.** Firebox door and back of boiler **19.** Bottom of chimney showing spark arrestor **20.** Locomotive reverser winding handle **21.** Vacuum brake ejector and steam brake control **22.** Vacuum brake gauge (left) and steam chest gauge (right) **23.** Speedometer **24.** Sand application valve **25.** Regulator handle (right) **26.** Automatic warning system (AWS) Sunflower dial **27.** Boiler gauge glass **28.** Steam supply control valve to air brake steam compressor (a modern addition) **29.** Tender water gauge indicator **30.** Fireman's side control valves (left to right): damper controls, ashpan and tender sprinkler valves, steam and water injector controls **31.** Tender coal bunker space holding about 7¼ tons (7.5 tonnes) of coal

LUCILLE PULLMAN CAR EXTERIOR

Built in 1928, *Lucille* started out as a first-class parlour car for the *Queen of Scots* Pullman service, then ran in the *Bournemouth Belle*. *Lucille* joined the *British Pullman* train in 1986. Fully restored to its former glory, the sides of the carriage proudly display its name and the Pullman coat of arms against the gleaming umber and cream livery.

1. Pullman coat of arms transfers on side **2.** Carriage name painted in gold lettering
3. Decorative gold embellishments on lower panels **4.** Decorative gold design on fascia embellishment **5.** Bogie **6.** Owner's plate **7.** Brass embarkation light above door
8. Elaborately designed door handle **9.** Passenger door into carriage **10.** Specification plate attached to end of carriage

LUCILLE PULLMAN CAR INTERIOR

A prominent feature of this 1920s-themed, first-class dining car is the French-polished wood panelling of the walls and partitions. The distinctive marquetry on the side panels was manufactured by Albert Dunn in the 1920s and features Grecian urns on dyed green holly wood. The panel restoration was completed by the Dunn family using the original green veneer. Other period touches include the Art Deco lampshades and brass fittings throughout. The windows beside each table are framed with curtains that add to the vintage feel. Luxurious, upholstered dining chairs seat up to 24 passengers, who are served by staff dressed in period-style uniforms.

11. *Lucille* carriage ready for service   **12.** Coat of arms and name plaque above doorway   **13.** Gold lettering of carriage name above door   **14.** Ceiling light with tulip glass shade **15.** Brass pull-down roof vent   **16.** Public address speaker grille **17.** Decorative, wooden marquetry panel featuring Grecian urn design   **18.** Wall-mounted brass "torch lights" with tulip glass shade   **19.** Embroidered motif on antimaccasars   **20.** Mirrors inlaid into wooden panelling   **21.** Emergency stop chain **22.** Brass individual seat number   **23.** Window latch **24.** Internal door handle   **25.** Four-seat private coupe **26.** Lavatory at end of carriage, accessed through corridor

CYGNUS PULLMAN CAR

Designed in 1938, *Cygnus*'s completion was delayed by World War II. Decorated with Australian walnut panelling, the car was reserved for use by travelling royalty and visiting heads of state. Along with *Perseus*, the car formed part of Sir Winston Churchill's funeral train in 1965.

1. Single-seat layout of car **2.** Pullman coat of arms on carpet at entrance to car **3.** Lavatory with mosaic floor depicting a swan **4.** Lavatory marble sink surround and wooden panelling **5.** Decorative-glass, cathedral light window in the lavatory

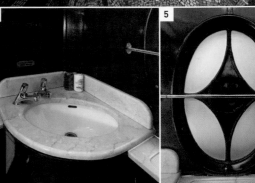

GWEN PULLMAN CAR

Originally on the iconic *Brighton Belle* service, *Gwen* is famous along with sister car *Mona* for conveying Queen Elizabeth (later the Queen Mother) to Brighton in 1948. After retirement *Gwen* was bought by VSOE in 1988, restored, and returned to the rails as part of the *Belmond British Pullman*.

6. Coat of arms and name plaque above doorway **7.** Decorative marquetry **8.** Mirrored wooden panels divide length of car **9.** View of *Gwen* car along central gangway **10.** Kitchen situated at end of car accessed through corridor

VERA PULLMAN CAR

Originally serving on the *Southern* and *Brighton Belles*, *Vera* was badly damaged during the 1940 blitz. It was returned to service seven years later when it was teamed up with *Audrey* as part of a five-car train. After retirement *Vera* was used as a summer house in Suffolk until the Venice Simplon-Orient Express (VSOE) bought it in 1985. *Vera* is renowned for its gleaming interior, decorated with sumptuous panels of sandalwood and mahogany. The dining chairs are upholstered in a copy of the original 1930s fabric.

11. Coat of arms and name plaque over doorway **12.** Wooden marquetry panelling **13.** Overview of *Vera* car **14.** Illuminated seat number **15.** Etched-glass sunburst wall light **16.** Detail of brass luggage rack **17.** Four-seater coupe

ZENA PULLMAN CAR

This car was built in 1929 as a first-class parlour car for the Great Western Railway's *Ocean Liner* services to Plymouth. *Zena* had an illustrious career and even hosted the French President Auriol on a state visit in 1950. The beautifully restored, sandalwood panels are inlaid with intricate motifs.

18. Coat of arms and name plaque above doorway **19.** Art Deco motif on wooden panelling **20.** Overview of *Zena* car ready for service

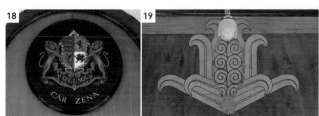

High Speed – The New Generation

By 2000, train speeds had increased significantly since the first high-speed Shinkansen and TGV trains of the 1960s to 1980s. New lines were being designed specifically for trains that could operate at 205 mph (330 km/h), and a new generation of trains was being introduced in several countries. Plans for intercity Maglev (*Magnetic Levi*tation) routes were developed in both Germany and Japan although, as yet, none has been built as the construction costs are too high. In the US 150-mph (241-km/h) operation on sections of existing lines was introduced.

△ **Trenitalia ETR 500, 2000**

Wheel arrangement 13-car trains including two power cars

Power supply 3,000 V DC, 25 kV AC, overhead lines

Power rating 11,796 hp (8,800 kW)

Top speed 211 mph (340 km/h)

These trains were based upon an earlier batch of ETR 500 trains built in the mid 1990s for operation on 3,000 V DC electrified lines of the Italian railways (Trenitalia). The new high-speed lines connecting Naples and Rome, and Florence and Milan that opened after 2000 needed trains able to work on 25 kV AC power, which the latest ETR 500 can do. The trains are limited to 186 mph (300 km/h) for current operation in Italy.

△ **DB ICE 3, 2000**

Wheel arrangement 8-coach, high-speed EMU

Power supply 15 kV AC, 16²/₃ Hz, 25 kV AC, 3,000 V DC, 1,500 kV DC, overhead lines

Power rating 10,724 hp (8,000 kW)

Top speed 205 mph (330 km/h)

Sixty-seven ICE 3 trains entered service from 2000, just before the new 205-mph (330-km/h) high-speed line connecting Cologne with Frankfurt airport opened in 2002. All had eight cars and featured "panorama lounges" at either end, where passengers could see the driver and the line ahead through a glass screen. Seventeen of the trains were four-voltage international sets, four of which were bought by Dutch Railways.

▷ **SMT/Transrapid, 2004**

Wheel arrangement Maglev (no wheels)

Power supply electromagnetic suspension

Power rating unknown

Top speed 268 mph (431 km/h)

The world's first commercial Maglev system was built at Birmingham Airport, UK in 1984. Work to develop high-speed Maglev systems was led by Japanese and German companies in the 1990s, and the world's only high-speed system opened in China in 2004 with a 19-mile (31-km) route connecting Shanghai city with its Pudong International Airport using German-built trains.

◁ **Chinese Railways CRH₂A, 2007**

Wheel arrangement 8-coach, high-speed EMU

Power supply 25 kV AC, overhead lines

Power rating 6,434 hp (4,800 kW)

Top speed 155 mph (250 km/h)

The Chinese Government ordered 60 CRH₂A trains from Kawasaki of Japan working with China Southern Rolling Stock Corp. (CSR) in 2004. The train is based on the E2 Shinkansen operated by Japan Railways (JR) East. The first three were built in Japan; the remainder were assembled at CSR Sifang. CSR has built several more variants since 2008 including 16-car sleepers.

△ Amtrak Acela, 2000

Wheel arrangement two Bo-Bo power cars
plus 6 passenger cars

Power supply 11 kV AC 25 Hz, 11 kV AC
60 Hz, and 25 kV AC 60 Hz, overhead lines

Power rating 12,337 hp (9,200 kW)

Top speed 165 mph (266 km/h)

Amtrak ordered the new Acela design following
trials of several European high-speed trains in
the US in the 1990s. Built to unique US standards
for crashworthiness, the trains can tilt, enabling
higher speed on curves. Current maximum speed
is 150 mph (241 km/h) in service, but plans exist for
160-mph (257-km/h) operation on some sections
of the Washington DC–Boston route in the future.

▽ RZD *Sapsan*, 2009

Wheel arrangement 10-coach, high-speed EMU

Power supply 25 kV AC, 3,000 V DC,
overhead lines

Power rating 10,728 hp (8,000 kW)

Top speed 155 mph (250 km/h)

German train manufacturer Siemens built eight 10-car
broad-gauge (5-ft/1.52-m) versions of its Velaro high-
speed train for Russia in 2009–11. Russian Railways
(RZD) operate the trains branded *Sapsan* (peregrine
falcon) – the fastest bird – between Moscow and St
Petersburg/Nizhny Novgorod. Eight more trains are
due to enter service in 2014–15.

△ JR N700 Shinkansen, 2007

Wheel arrangement 16-coach,
high-speed EMU

Power supply 25 KV AC, overhead lines

Power rating 22,905 hp (17,080 kW)

Top speed 186 mph (300 km/h)

The N700 Shinkansen can accelerate faster
than any of the trains it replaced on the
Tôkaidô Shinkansen line between Tokyo and
Hakata. Built in either 8- or 16-car train sets,
most N700s were in service by 2012, and 149
trains will have been delivered by the time
production ends in 2016.

TALKING POINT

Meals on the Move

"Bento" is the Japanese name for a carefully crafted takeaway meal in a
single, often disposable, container. Bento boxes were historically made from
wood or metal, but are now found in a variety of materials and novelty shapes.
A wide range of bento box train meals, known as ekiben, are sold at kiosks
in railway stations all over Japan to take on board the train.

Novelty boxes
Ekiben packed
in boxes shaped
like Japanese
trains have
become collector's
items. This one is
modelled on the
N700 Shinkansen.

Spectacular Stations

Railway stations have become sites of some of the world's most exquisite architecture and design. Whether the look is contemporary or classic, the architecture of the most celebrated stations successfully combines form and function to make an indelible impression on every traveller who passes through them.

1. **The Tsuzumi Gate of Kanazawa Station in Tokyo** (2005) combines traditional Shinto temple designs with the strings of a Japanese drum.
2. **Danggogae Station in Seoul** (1993) is the north terminus of Line 4 of the South Korean capital's subway.   3. **Financial Centre metro station in Dubai** (2009) was designed with a shell-like structure that recalls the city's early prosperity from pearl diving.   4. **Komsomolskaya Station in Moscow** (1952) is located on the metro's Koltsevaya Line. Its grand architecture features chandelier lighting, baroque details, and mosaics of historic Russian scenes.
5. **Haydarpasa Terminus in Istanbul** (1908) has neoclassical styling, is surrounded on three sides by water, and is Eastern Europe's busiest station.
6. **Tanggula Station in Tibet** (2006) is the highest railway station in the world at 16,627 ft (5,068 m) above sea level.   7. **Chhatrapati Shivaji Terminus in Mumbai** (1888) has Gothic-style architecture and has been recognized by UNESCO as a world heritage site.   8. **Berlin Hauptbahnof** (2006) is a glass and steel structure with platforms on two levels, serving 350,000 passengers daily.   9. **St Pancras International in London** (1868) reopened in 2007 following extensive renovation and expansion, but retained the original Victorian roof to dramatic effect.   10. **Liége-Guillemins Railway Station in Liége** (2009) has no outer walls, but a glass and steel canopy that covers all five platforms. The station serves as a transport hub for high-speed rail links across Europe.   11. **Grand Central Terminal in New York** (1913) has 44 platforms – more than any other station in the world. The Beaux-Arts architecture features Botticino marble staircases and an astronomical ceiling.
12. **Union Station in Los Angeles** (1939) has the appearance of a church from the outside, but is one of the busiest stations on the US west coast, serving more than 60,000 passengers daily.   13. **Ushuaia Station in Tierra Del Fuego** (1910) was originally used to transport prisoners to an Argentine penal colony. It closed in 1947, but reopened in 1994 after extensive renovation; the station has since become a popular tourist attraction.

Faster and Faster

Many existing high-speed train fleets have been expanded and, as new lines have opened, increasing numbers of fast international services have become possible, connecting France, Germany, Spain, and Switzerland in particular. In France, a specially modified test train built by Alstom, the TGV V150, achieved a new world speed record of 357¼ mph (574.8 km/h) in 2007. In Japan the main railways had been privatized by 2006, and in 2012 the world's first "start-up" private high-speed operator was NTV in Italy, which also began its services with a brand-new train design. However, state-owned operators continue to dominate.

△ NTV AGV ETR 575, 2012

Wheel arrangement 11-coach, articulated, high-speed EMU

Power supply 3,000 V DC, 25 kV AC, overhead lines

Power rating 10,054 hp (7,500 kW)

Top speed 186 mph (300 km/h)

Italian private rail operator Nuovo Trasporto Viaggiatori (New Passenger Transport), started high-speed services from Naples to Rome and Turin in 2012 with a fleet of 25 Alstom-built Automotrice à Grande Vitesse (AGV) high-speed trains. The trains are equipped with three different classes of passenger accommodation.

△ VT Class 390 Pendolino, 2002

Wheel arrangement 9- or 11-coach, high-speed, tilting EMU

Power supply 25 kV AC, overhead lines

Power rating 7,979 hp (5,950 kW)

Top speed 124 mph (200 km/h)

The Virgin Trains's tilting, high-speed Pendolino trains have speeded up journeys on the UK's West Coast Main Line from London to Birmingham, Manchester, and Glasgow since 2002. By leaning on curves the trains can go faster than conventional ones, reducing journey times and increasing track capacity.

Steam Train Revival

The UK has led the world in preserving mainline steam locomotives, with many restored to working order since the 1950s. The success in preserving different types has led to groups of volunteers trying to recreate engine classes that never survived. All 49 Peppercorn Class A1 locomotives were scrapped in the 1960s, so in 1990 a group decided to build another – from scratch. Nineteen years later the new engine, based on the original design but with some modern features, started operations. It now works charter trains all over the country. Several other similar projects are now underway in the UK.

Peppercorn Class A1 No. 60163 _Tornado_, 2008
This is the first main-line steam engine built in the UK since 1960.

▽ LSER Class 395 Javelin, 2009

Wheel arrangement 6-car EMU

Power supply 25 kV AC overhead lines and 750 V DC third rail

Power rating 4,506 hp (3,360 kW)

Top speed 140 mph (225 km/h)

Built by Hitachi in Kasado, Japan, using Shinkansen technology, the Javelin trains have been in service since 2009 serving the London & South Eastern Railway on the UK's domestic high-speed line HS1. They have reduced journey times significantly (in some cases by as much as than 50 per cent) between cities in Kent and London.

△ PKP IC Class ED250, 2014

Wheel arrangement 7-coach, high-speed EMU

Power supply 15 kV AC, 16²/₃ Hz, 25 kV AC, 3,000 V DC, 1,500 kV DC, overhead lines

Power rating 7,373 hp (5,500 kW)

Top speed 154 mph (249 km/h)

Poland's long-distance railway company Polskie Koleje Panstwowe Intercity (PKP IC) ordered 20 Class ED250 trains from Alstom. Based on the "New Pendolino" design, first built for China and Italy, but without the tilt equipment, the trains began to replace older locomotive-operated trains on the Gdynia/Gdansk–Warsaw–Krakow/Katowice route from late 2014.

△ SNCF TGV POS, 2006

Wheel arrangement 10-car train including 2 power cars

Power supply 15 kV AC, 16²/₃ Hz, 25 kV AC, 1,500 kV DC, overhead lines

Power rating 12,440 hp (9,280 kW)

Top speed 199 mph (320 km/h)

Using the new LGV Est high-speed line that connects the Alsace region with Paris, the French national railways' (SCNF) TGV POS (Paris–Ostfrankreich–Süddeutschland, or Paris–Eastern France–Southern Germany) trains started operating from 2006. The TGV POS trains also work through services to Munich and Frankfurt in Germany plus Geneva and Zurich in Switzerland.

◁ SNCF TGV Euroduplex, 2012

Wheel arrangement 10-car train including 2 power cars

Power supply 15 kV AC, 16²/₃ Hz, 25 kV AC, 1,500 kV DC, overhead lines

Power rating 12,440 hp (9,280 kW)

Top speed 199 mph (320 km/h)

The third-generation "Duplex" (double-deck) TGV train was built for the French national railway SNCF; 95 have been ordered, with deliveries planned until 2017. This is the only double-deck high-speed train capable of operating across several different European rail networks and is used for services between France and Germany and France and Spain.

▷ SNCF TGV V150, 2007

Wheel arrangement 5-car, TGV train

Power supply 31 kV AC, overhead lines

Power rating 26,284 hp (19,600 kW)

Top speed 357¹/₄ mph (574.8 km/h)

This test train used two new TGV POS power cars and three special cars with powered bogies, all vehicles having specially made, bigger wheels. For the test run on the LGV Est Européenne line, the overhead line voltage was increased for more power. The record of 357¹/₄ mph (574.8 km/h) – 6 miles (9.65 km) per minute – more than achieved the target set.

Javelin No. 395 017

Built by Japanese manufacturer Hitachi, the Class 395 Javelin is based on technology used in Shinkansen trains. A multiple-unit train set with a power car at each end, the Javelin can be powered by overhead wires, but uses a third-rail electricity pick-up when operating on conventional lines in southeast England. Javelin trains have reduced some journey times by as much as 50 per cent, and they provided a key service during London's 2012 Summer Olympic Games.

THE HIGH-SPEED LINE from London to the Channel Tunnel, known as High Speed 1 (HS1), was finished in November 2007. Providing international services via the tunnel, it also allows domestic high-speed Class 395 Javelins to serve southeast England. Although slower than the 189-mph (304-km/h) Eurostar trains with which it shares the HS1, the lighter, shorter Javelin accelerates faster.

The Javelin has four safety systems. Two French systems are used on the HS1; Transmission Voie Machine (TVM430) sends signalling information through the track to displays in the driver's cab, and Contrôle Vitesse par Balise (KVB) monitors and controls the train's speed from St Pancras International station, London. On conventional British routes, the Automatic Warning System (AWS) and Train Protection and Warning System (TPWS) work together to alert the driver to signals, and will stop a train if it passes a danger signal.

FRONT VIEW (DPT2)

FRONT VIEW (DPT1) - EXTRA YELLOW PANELS INDICATE UNIVERSAL ACCESS

SPECIFICATIONS			
Class	395	In-service period	2009-present
Wheel arrangement	6-car EMU	Railway	Southeastern
Origin	Japan	Power supply	25 kV AC overhead wires and 750 V DC third rail
Designer/builder	Hitachi, at Kasado	Power rating	4,506 hp (3,360 kW)
Number produced	29 Class 395	Top speed	140 mph (225 km/h)

High-speed designer
German designer Alexander Neumeister created the look of the Javelin, which bears a resemblance to his German ICE 3 and several Japanese Shinkansen variants. Neumeister has also designed Chinese and Russian high-speed trains.

Sliding doors are automated

Safety line warns rail staff of the presence of overhead power cables

Electrical equipment including heating and air-conditioning units

Name honours one of 24 British athletes from the 2012 London Olympics

Third-rail shoe picks up electrical power from third rail

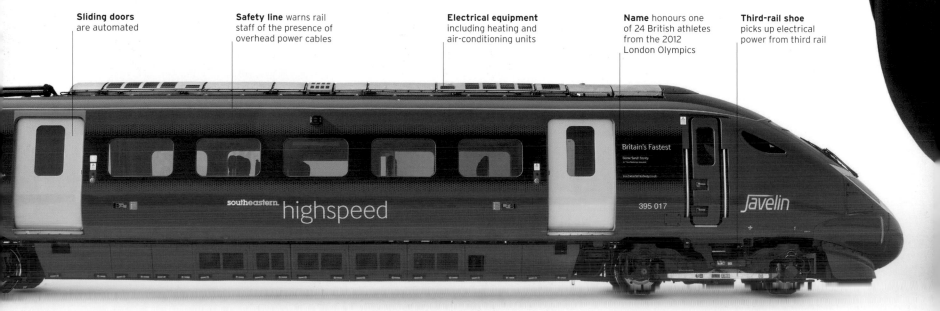

Javelin

Serving the Olympic Games

Ferrying millions of visitors to the Olympic Park via HS1 and the station at Stratford International, the 2012 Olympic shuttle was known as the Javelin, and the name endures today.

EXTERIOR

The aluminium car bodies are adorned in a dark blue livery that is unique to the Javelin trains. Each car has two wide single sliding doors on each side, painted in lighter colours to be easily identified by passengers. The driving cars at each end hold the pantographs and third-rail collector shoes, which pick up electricity to power the train.

1. Olympian signature on side of front vehicle   **2.** Headlight (above) and tail light (below)   **3.** Coupler and horn inside open nose cone   **4.** External emergency door release (access handle)   **5.** Driving cab door handle   **6.** Driver's cab door   **7.** Southeastern logo on side   **8.** Rheostatic brake resistor mounted on roof   **9.** Pantograph assembly   **10.** Vacuum circuit breaker (VCB)   **11.** Third-rail shoe fuse   **12.** Axle end earth on rear vehicle   **13.** Underframe view of motor bogie

Sarah Storey

Britain's Fastest

Dame Sarah Storey

22 Time Paralympic Medallist

southeasternrailway.co.uk

CAB INTERIOR

The driver's cab is typical of those found in modern high-speed trains, with a single driver's chair positioned centrally facing the control desk. To the right are CCTV displays from the passenger cars and to the left is the Train Management System (TMS), which, among other functions, allows the driver to switch between third rail and overhead power.

14. Driver's seat and controls   **15.** Train management system (TMS)   **16.** Combined power/brake controller (CPBC)   **17.** CCTV panel position in driving cab **18.** Emergency brake push button   **19.** Master key **20.** Secondary seat in driver's cab   **21.** Short circuiting bar and red flag, for use in emergencies   **22.** Miniature circuit breaker (MCB) panel   **23.** Switch panel on cab back wall **24.** Onboard manager's door control panel in driving cab

CARRIAGE INTERIORS

The Class 395 is designed for commuter journeys
that typically last no more than an hour. The train
does not offer first-class accommodation, but its
passengers pay higher fares than they would on
the slower, conventional trains. The gangway runs
through all six carriages and connects them, but if
two trains are joined together, it is not possible to
get from one six-car section to the other.

The interior of the carriages is blue and grey,
and complements the dark blue external livery.
Each train contains 340 2+2 seats, with the
majority of them facing in the same direction. In
addition, there are 12 tip-up seats in the door
vestibule areas. Two toilets are provided per train,
one of which is designed for universal access.
Passenger information interfaces on the train
include digital displays and a PA system.

25. Overview of carriage **26.** Seats with table **27.** Luggage
rack above seats **28.** Adjustable armrest **29.** Tray table
on seat back **30.** Interior luggage area **31.** Gangway
door between carriages **32.** Gangway door open button
33. Passenger information system (PIS) display panel
34. Disabled toilet **35.** Power socket sign above seats
36. Emergency alarm sign **37.** Hand hold on seat for
passengers walking in the aisle **38.** Interior of disabled
carriage **39.** Passenger door open/close buttons

Dubai Metro

Worsening traffic congestion and a growing population – expected to reach 3 million by 2017 – persuaded Dubai's leaders to build the United Arab Emirates' first metro. In May 2005 a US $3.4 billion contract was awarded to Dubai Rail Link (DURL), a consortium of companies from France, Japan, Turkey, and the US. Two routes were planned: the Red Line and the shorter Green Line. Work began in 2006 on the Red Line, with stations serving the city centre.

The opening of the first section on 9 September 2009 attracted more than 110,000 people. The five-car trains could carry up to 643 passengers and provided three classes of travel, including a section for women and children. However, the most impressive aspect was that the trains were fully automated. This was the inaugural stage of what – at 47 miles (75 km) – was in 2012 declared to be the world's longest driverless rail network.

PLANS FOR EXPANSION

The Red Line fully opened in April 2010, by which time the inaugural section had already carried more than 11 million passengers. The second route, the Green Line, was opened in September 2011. The Metro's success has led to plans to extend the existing lines and create three addtional routes which, by 2030, would enlarge the Dubai Metro to 262 miles (421 km) of track, servicing a total of 197 stations.

The driverless trains on the Red Line Metro in Dubai pass over viaducts above the streets as well as underground. Power is drawn from a third rail.

Into the Future

Investment in the railways around the world is growing, driven by rising passenger numbers as large cities continue to expand, combined with increasing road congestion, and the need to reduce CO_2 emissions. Older-style trains that use locomotives and separate coaches are being replaced by modern, self-powered multiple units. Rail operators, both passenger and freight, are also seeking to reduce maintenance and energy costs; some modern trains are designed to recycle electricity while braking.

◁ **Bombardier Omneo Régio2N, 2010**

Wheel arrangement	6- to 10-car, articulated EMU
Power supply	25 kV AC, 1,500 kV DC, overhead lines
Power rating	4,291 hp (3,200 kW)
Top speed	99 mph (160 km/h)

The Omneo is the world's first articulated, double-deck EMU with a single-deck driving coach at each end, and double-deck, articulated intermediate coaches sandwiched between short, single-deck door sections. The trains can be supplied in lengths ranging from 6 to 10 cars (266-443 ft/81-135 m). The French national railway (SNCF) has agreed a €7-billion-framework contract for up to 860 trains for delivery until 2025.

△ **Bombardier Zefiro 380, 2012**

Wheel arrangement	8-car EMU
Power supply	25 kV AC, overhead lines
Power rating	13,454 hp (10,037 kW)
Top speed	236 mph (380 km/h)

The latest version of the Bombardier-designed Zefiro high-speed train is for operation at up to 236 mph (380 km/h). Chinese Railways have ordered 70 (two were delivered in 2012). In Europe 50 224-mph (360-km/h) versions are being built for Italian operator Trenitalia, and enter service from 2014.

Local Transport Developments

The demand for urban transport has grown significantly in the last 30 years – whether metros under city streets or light-rail systems that run on roads alongside other vehicles. The strongest growth is in Asia and the Middle East where new systems have been built since 1990. For established networks the challenge is to create more capacity through better performance and smart control systems on networks that are more than 100 years old, for example, in London and Paris.

▷ **Vossloh Wuppertal Schwebebahn train, 2015**

Wheel arrangement	3-section, articulated vehicle
Power supply	750 V DC, third rail adjacent to single running rail
Power rating	322 hp (240 kW)
Top speed	approx. 37 mph (60 km/h)

Germany's Wuppertal Schwebebahn is a suspended railway built largely above the River Wupper on massive iron supports. First opened in 1901, it is now a protected national monument, but is still used daily by thousands of commuters. Vossloh will supply 31 new trains from 2015 – part of a comprehensive modernization plan.

◁ **Amtrak Siemens American Cities Sprinter ACS-64, 2014**

Wheel arrangement Bo-Bo

Power supply 25 kV, 12.5 kV, and 12 kV AC, overhead lines

Power rating 8,579 hp (6,400 kW)

Top speed 125 mph (201 km/h)

Siemens is building 70 ACS-64s at its factory in Sacramento, California. Amtrak, which introduced the first ACS-64 in 2014, will use them to replace all its existing electrics on the Washington DC-New York-Boston Northeast Corridor route.

△ **VMS Chemnitz tram-train, 2015**

Wheel arrangement 3-section articulated LRV

Power supply 600 V and 750 V DC, overhead lines plus diesel engines

Power rating electric: 777 hp (580 kW); diesel: 1,046 hp (780 kW)

Top speed 62 mph (100 km/h)

Tram-trains that enable travel to city centres from regional railway lines are now in use in many EU countries. In Germany some use diesel engines on non-electrified rail lines. Chemnitz tram-trains will use this technology from 2015.

TECHNOLOGY
Cargo Efficiency

The major Class 1 Railways in North America have increased operational efficiency and productivity significantly since the 1980s. By operating longer, heavier trains using powerful modern locomotives, operating costs per cargo container have reduced, making rail much cheaper than road. Double-stacked containers are used in North America, Australia, and India. The Brazilian mining company Vale runs the 554-mile (892-km) Carajás Railroad with the world's heaviest trains – 330-wagon, 41,632-ton (42,300-tonne) iron-ore trains run up to 24 times a day to the port at Ponta da Madeira.

BNSF freight train, Cajon Pass, California With two modern GE Evolution Series ES44DC engines at each end, this train can be up to $2^2/_3$-mile (4.3-km) long. On steep gradients, the engines slow down descending trains, as well as pull them up the inclines.

△ **Siemens Vectron, 2013**

Wheel arrangement Bo-Bo

Power supply 3 kV DC, overhead lines

Power rating 6,974 hp (5,200 kW)

Top speed 99 mph (160 km/h)

Siemens developed the Vectron family of locomotives to replace its previous Eurosprinter model. The first major order received was for 23 Vectron DC electric locomotives from Polish rail freight operator DB Schenker Rail Polska – the first of these entered service in 2013. Subsequent orders for locomotives for use in several countries have been obtained, including a broad-gauge version for Finland.

△ **Siemens ICx, 2017**

Wheel arrangement 7- or 12-coach, high-speed EMU

Power supply 15 kV AC, $16^2/_3$ Hz

Power rating 13,271 hp (9,900 kW)

Top speed 155 mph (250 km/h)

ICx trains will replace Germany's existing long-distance, locomotive-operated trains, and later the first two types of ICE train. Due for delivery from 2017 are 85 12- and 45 slower 7-coach trains using 92-ft (28-m) long coaches configured as distributed-power EMUs with more seats and space than those they replace.

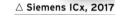

△ **Calgary Transit C-train System Siemens S200, 2015**

Wheel arrangement 2-car, articulated LRV

Power supply 600 V DC, overhead lines

Power rating 777 hp (580 kW)

Top speed 65 mph (105 km/h)

The Canadian city of Calgary opened its first light-rail line in 1981. Since then the network has expanded and carries 290,000 people daily. To increase capacity and to retire some of the original light-rail vehicles (LRVs), 60 new S200 LRVs are on order for delivery in 2015-16. Calgary Transit expects to increase its fleet from under 200 to 390 over the next 30 years.

△ **London Underground Siemens Inspiro metro concept**

Wheel arrangement 6-car, metro EMU

Power supply 630 V DC, third and fourth rail

Power rating 1,340 hp (1,000 kW)

Top speed 56 mph (90 km/h)

London Underground has seen significant growth in passengers. The "New Tube for London" programme is planning 250 new underground trains, possibly automatic and driverless, to enter service between 2020 and 2035. Three companies are designing trains; shown here is Siemens's proposal.

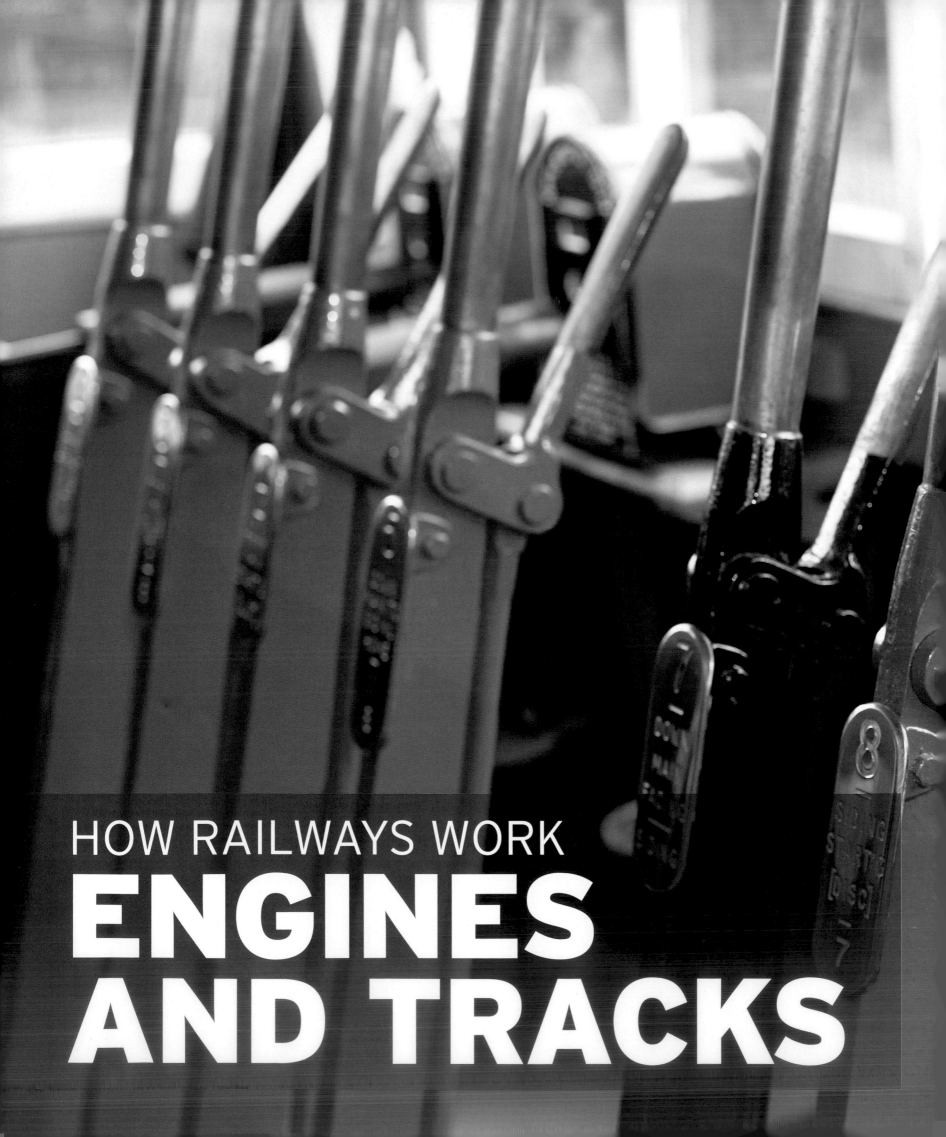

ENGINES AND TRACKS

How Tracks Work

Wooden rails were used for the pony-drawn wagonways of the 17th century, but iron was required to support the steam engines of the 19th century. The cast-iron rails of the first railways were succeeded by sturdier, wrought-iron rails in the 1820s, before steel – stronger still – came into use. Steel rails were first laid at Derby Station in England in 1857. The rails, sleepers, and ballast of a finished railway line have come to be known as the "permanent way", a term that dates back to the earliest days of railway construction when a temporary track was laid first in order to transport materials to where the line was being constructed. The temporary track was replaced by the permanent way once the substructure was largely completed. The "gauge" – the distance between the rails – and the alignment of the rails are constantly monitored during construction to ensure that they remain uniform throughout the straight sections and curves in the track.

TRACK FORMATION

The substructure of a track is called the "formation". Since a consistent "grade" (gradient) is required for trains to run smoothly, the ground is first prepared to form the "subgrade". The subgrade might also be covered by a layer of sand or stone called a "blanket" before it is overlaid with ballast. Sleepers are bedded into the ballast to support the rails. Crushed-stone ballast is still the most common foundation and allows for good drainage.

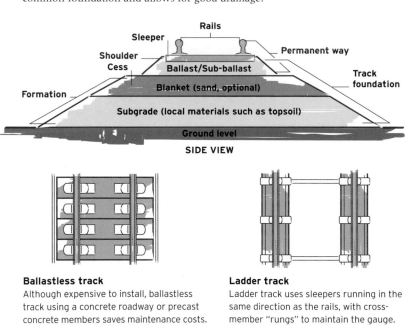

Ballastless track
Although expensive to install, ballastless track using a concrete roadway or precast concrete members saves maintenance costs.

Ladder track
Ladder track uses sleepers running in the same direction as the rails, with cross-member "rungs" to maintain the gauge.

TRACK STRUCTURE

Most modern railway tracks consist of flat-bottom steel rails fixed to timber or concrete sleepers. Flat-bottom steel rails are more stable, easier to lay, and do not suffer from wear in the same way as the old cast-iron or wrought-iron rails. Bull-head rails are the same shape top and bottom so that they can be turned over and reused when the head becomes worn.

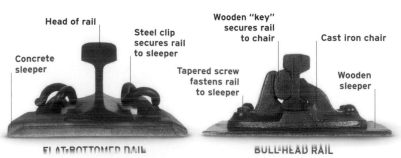

FLAT-BOTTOMED RAIL

BULL-HEAD RAIL

TRACK GAUGE

The gauge of a railway's tracks is defined as the distance between the rails, measured from the inside of the rail – except in Italy, where it can be the distance between the centre of the rails (see below). The first railway builders chose whatever gauge they felt was appropriate for their line; a wider gauge was thought to give greater stability for a train at speed or in strong crosswinds, while a narrow gauge took up less space and was usually cheaper to build. When lines grew into networks, some form of standardization became essential.

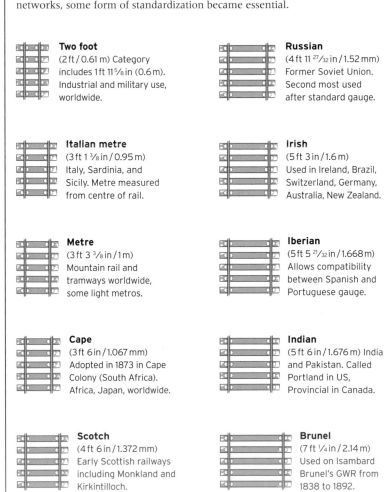

Two foot
(2 ft / 0.61 m) Category includes 1 ft 11⅝ in (0.6 m). Industrial and military use, worldwide.

Russian
(4 ft 11 ²⁷/₃₂ in / 1.52 mm) Former Soviet Union. Second most used after standard gauge.

Italian metre
(3 ft 1 ⅜ in / 0.95 m) Italy, Sardinia, and Sicily. Metre measured from centre of rail.

Irish
(5 ft 3 in / 1.6 m) Used in Ireland, Brazil, Switzerland, Germany, Australia, New Zealand.

Metre
(3 ft 3 ⅜ in / 1 m) Mountain rail and tramways worldwide, some light metros.

Iberian
(5 ft 5 ²¹/₃₂ in / 1.668 m) Allows compatibility between Spanish and Portuguese gauge.

Cape
(3 ft 6 in / 1.067 mm) Adopted in 1873 in Cape Colony (South Africa). Africa, Japan, worldwide.

Indian
(5 ft 6 in / 1.676 m) India and Pakistan. Called Portland in US, Provincial in Canada.

Scotch
(4 ft 6 in / 1.372 mm) Early Scottish railways including Monkland and Kirkintilloch.

Brunel
(7 ft ¼ in / 2.14 m) Used on Isambard Brunel's GWR from 1838 to 1892.

Standard
(4 ft 8 ½ in / 1.435 m) Used on 60 per cent of the world's railways including US and UK.

Breitspurbahn
(9 ft 10 in / 3 m) Proposed for Hitler's Third Reich supertrain but railways never built.

How Wheels Work

The wheels of a train are designed to enable it to follow curves in the track. Each wheel tapers from the inside outwards and has a projecting flange on the outer edge. The flange is to prevent the wheels from derailing and normally it never comes into contact with the track, the weight of the train being borne by the conical surface. These sloping edges allow the wheels to slide across the tops (heads) of the rails. The wheels on the outside of a curve have further to travel, so use the larger radius close to the flange, while the wheels on the inside use the shorter radius closest to their centre.

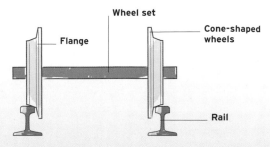

The flanged wheel
The flanged wheel was invented by English engineer William Jessop in 1789 to provide a better grip on railed track; this helped to prevent derailments.

STEAM LOCOMOTIVE WHEEL CONFIGURATION

As steam locomotives grew bigger and heavier, they gained more wheels, spreading their weight more evenly and giving better traction. To describe the wheel arrangement, mechanical engineer Fredcrick M. Whyte came up with a numbering system in 1900. A locomotive with four leading wheels, four powered wheels, and two trailing wheels was a 4-4-2. Articulated locomotives, designed to tackle bends more easily, needed longer numbers, but the codes retained their simple logic. The Whyte system is used in the US and the UK for steam engines, although different systems are used elsewhere, and for other types of locomotive. Some configurations also had names.

WHEELS	TYPE	NAME
	0-2-2	Northumbrian
	2-2-0	Planet
	2-2-2	Jenny Lind or Patentee
	4-2-0	One Armed Billy
	0-4-0	Four-wheeler
	4-4-0	American or Eight-wheeler
	4-4-2	Atlantic
	2-6-0	Mogul
	2-6-2	Prairie
	4-6-0	Ten-wheeler or Grange
	4-6-2	Pacific
	4-6-4	Baltic or Hudson
	2-8-0	Consolidation
	2-8-2	Mikado, Mike, or MacArthur
	2-8-4	Berkshire
	4-8-4	Northern
	2-10-4	Texas or Selkirk
	0-4-4-0	*none*
	2-6-6-2	*none*
	2-6-6-6	Challenger
	4-6-6-4	Blue Ridge or Allegheny
	2-8-8-4	Yellowstone
	4-6-4+4-6-4	Double Baltic

BRAKES

The first train brakes worked like the brakes on a horse-drawn cart, with levers moving a brake shoe to press a wooden block against the wheel tread. This was not very efficient and caused wear. Modern trains use disc brakes like those fitted to cars. The discs are attached to the axles, and calipers fitted with composite brake blocks "pinch" the discs to slow the train.

RIM BRAKE   DISC BRAKE

AIR BRAKE
During the 1870s two different types of brake systems were tried a vacuum system and an air brake system. Air brakes were shown to bring a 15-car train travelling at 50 mph (80 km/h) to a halt in half the time taken by vacuum brakes. The braking distance for the Westinghouse air brake was 777 ft (237 m), while a vacuum brake took 1,477 ft (450 m). Air or pneumatic braking is the standard system used today by the world's railways.

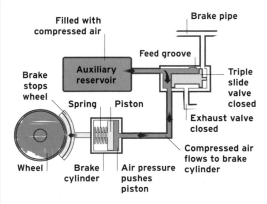

Air brake application
A pump compresses air for use in the system. The driver controls the air with a triple valve. When this is applied, compressed air is released into the brake pipe and air pressure forces the piston to move against a spring in the brake cylinder, causing the brake blocks to make contact with the wheels.

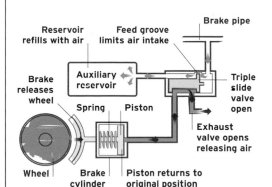

Air brake release
When the driver releases the brake valve, air leaves the brake pipes. As air escapes from the exhaust, a spring in the brake cylinder pushes the piston back, causing the brake blocks to disengage from the wheels. The auxiliary air reservoir, meanwhile, refills.

How Signals Work

In the earliest days of the railways, there were few trains and no real need for signalling systems. Trains ran up and down single tracks, and timetables kept them far enough apart to avoid accidents. As railway networks became more extensive, with rail traffic travelling from far and wide, timetables based on local time (most nations did not have standard time until the late 19th century) caused huge confusion, and signalling became essential to prevent collisions. By the 1830s the hand and lamp signals used by rail staff were being imitated by more visible mechanical trackside signals, although in some countries it would take almost a century for the style of these signals to be standardized across different networks.

Signal lamp
Lamps used by train guards, brake men, or station staff had different lenses that could shine a red, green, or clear white light.

EARLY SIGNAL SYSTEMS

The first trackside signals came in a variety of guises but, like the signalling lamps employed before them, used the colour red to mean "stop". Long recognized as the international colour for danger, red was an obvious choice. Green for "go" had also been used in lamps and was chosen as it couldn't easily be mistaken for red, or for a non-signal light that a locomotive driver might happen to see. Yellow lights, the colour adopted because it was distinct from the other two, were later introduced to advise caution. The trackside semaphore style of signal became the most widespread type and is still in use today.

BALL TOKENS USED ON INDIAN RAILWAYS

Signalling tokens
Tokens were used on single lines to ensure only one train could enter each "block" (section) of the line at a time. The crew collected a token from a signalman when it was safe to enter the block, handing it to another signalman at the end of the block. Automated systems were later developed with tokens dispensed and recovered by machines.

Ball signal
The most common signal on the early US railways, the ball signal gave rise to the term "highball". When it was raised, it was safe to proceed. This was later reversed.

Semaphore
Widespread after the 1850s, and still in use today, semaphore arms signalled "danger" when in the horizontal position and "all clear" when angled either up or down.

Wood's crossbar signal
Crossbar signals, in use from the 1830s, indicated on/off (stop/go) with a revolving wooden board. When the crossbar was swung parallel to the line it signalled clear.

Revolving disc signal
The disc revolved vertically to signal stop and go, much like semaphore signals. In keeping with most signals of the time, the disc was made of wood and painted red.

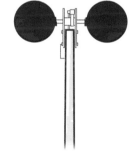

Double disc signal
Like the crossbar, the double disc rotated on a wooden or steel signal post. Both were short-lived, however, as the "clear" signal was hard for train drivers to see.

ELECTRIC LIGHT SIGNALS

The use of electric signal lights instead of oil lamps started to become common in the 1920s, although early electric lights were still not powerful enough for drivers to see them clearly from a safe distance in daylight. Semaphore signals were far clearer and easier to spot. It was not until 1944 that modern lenses improved the visibility of electric signals sufficiently to allow them to replace semaphore signals fully. Railway light signals have red lights at the bottom so that they are in the driver's line of sight; road traffic lights have red lights at the top so that drivers can see them above other cars.

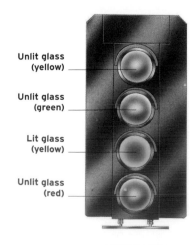

Unlit glass (yellow)

Unlit glass (green)

Lit glass (yellow)

Unlit glass (red)

FRONT VIEW

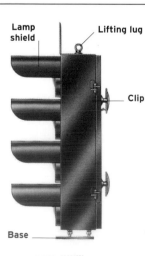

Lamp shield

Lifting lug

Clip

Base

SIDE VIEW

Stopping the train
Signal lights control their own block of track and are designed to give a driver all the warning he needs to slow down before reaching a hazard. Modern rail networks also have safety systems in place to apply a locomotive's brakes automatically if it passes a danger signal.

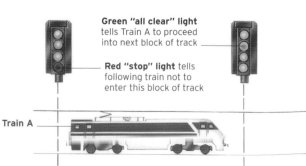

Green "all clear" light tells Train A to proceed into next block of track

Red "stop" light tells following train not to enter this block of track

Train A

SEMAPHORE SIGNAL

Semaphore signal systems began to be introduced in the 1840s and consisted of two pivoting arms or "blades" and a "spectacle" holding two coloured glass lenses. As the arm moved, the lenses moved in front of a light source, initially an oil lamp, to allow the signals to be seen at night. On the top arm the lenses were red and blue, the blue combining with the yellowy flame of the oil lamp to make green. On the bottom arm the lenses were yellow and blue. Once electric lights were used, the blue lenses were changed to green.

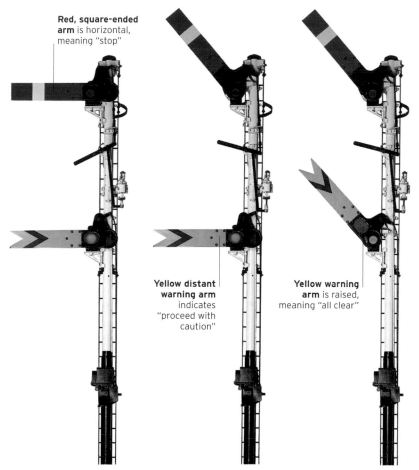

Red, square-ended arm is horizontal, meaning "stop"

Yellow distant warning arm indicates "proceed with caution"

Yellow warning arm is raised, meaning "all clear"

Stop
When the upper arm is horizontal, it means stop. The lower arm is a "distant" warning, telling the driver that the train may have to stop at the next signal. Both arms horizontal means stop.

Proceed with caution
With the upper "stop" arm raised and its light green, the train can proceed, but the lower distant warning arm is still telling the driver to be cautious as the next signal may require the train to stop.

All clear
When both the upper and lower arms are raised and both lights show green, it means that the line ahead is clear. The driver can proceed safely at normal speed until they arrive at the next signal.

POINTS

The points mechanism is a key component of any railway. Invented by English engineer Charles Fox in 1832, the simplest system used a pull rod activated by a lever. This adjusted movable track sections (points) to direct a train onto a curve, taking it away from the main line. Switching the points was a task performed by a signalman, but most points are now operated electronically.

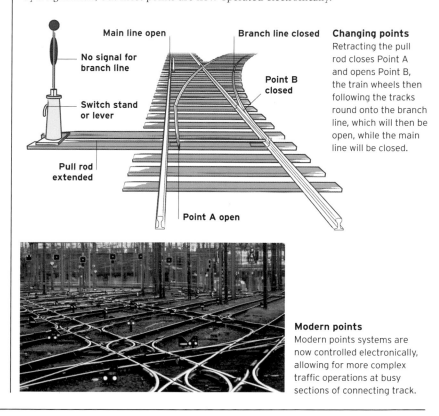

Main line open

No signal for branch line

Switch stand or lever

Pull rod extended

Point A open

Branch line closed

Point B closed

Changing points
Retracting the pull rod closes Point A and opens Point B, the train wheels then following the tracks round onto the branch line, which will then be open, while the main line will be closed.

Modern points
Modern points systems are now controlled electronically, allowing for more complex traffic operations at busy sections of connecting track.

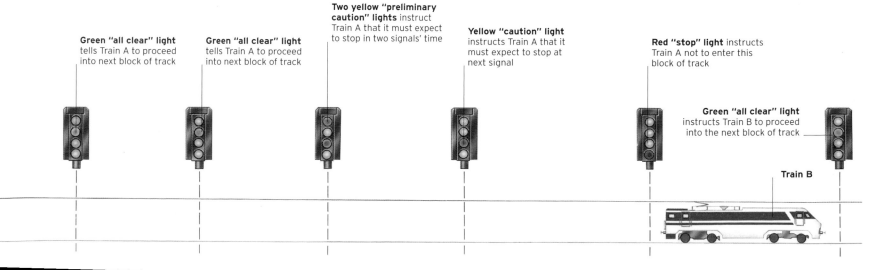

Green "all clear" light tells Train A to proceed into next block of track

Green "all clear" light tells Train A to proceed into next block of track

Two yellow "preliminary caution" lights instruct Train A that it must expect to stop in two signals' time

Yellow "caution" light instructs Train A that it must expect to stop at next signal

Red "stop" light instructs Train A not to enter this block of track

Green "all clear" light instructs Train B to proceed into the next block of track

Train B

Radstock North Signal Box

Before the days of automated signalling centres, signalmen managed the movement of trains from local signal boxes. The Radstock North signal box once controlled the trains on the old Great Western Railway North Somerset Line in the UK, and was restored at Didcot to represent an original box from the 1930s.

WHETHER IT WAS TO CONTROL a stop signal or to switch a passing train onto a different track, signal boxes once served as the control hubs of a rail transport system. The boxes ensured that trains operated safely over the correct route and in accordance with the scheduled timetable, and also provided the signalman with a warm and dry working environment. The earliest railway signals were given by hand, or with the issuing of tokens, but over time signalling became more mechanical, using levers housed in the signal box and positioned next to the track. The manually operated signal boxes were often raised to accommodate the movement of the lower part of the levers, and to allow the signalman a clear view of the surrounding track. Nowadays, with the advent of electronic signalling technology, traditional signal boxes have largely been replaced with centrally managed signalling control centres.

INTERIOR

The signal box contains numerous levers set inside a frame mounted on a beam beneath the floor. The levers are painted according to their function. The large wheel operates the level crossing, and the instruments on the shelf above the levers indicate whether or not different sections of the line are clear. The key token instrument offers a safety measure to ensure no two trains are ever on the same line on a collision path.

1. Overview of signal box interior **2.** Token equipment for branch line (left) and main line (right) **3.** Shelf containing block instruments and telegraph equipment **4.** Close-up of three-position block instruments **5.** Control levers: red for signals; blue for operating locks and gates; black for controlling points **6.** Large wheel and levers for operating level crossing **7.** Wicket gate levers **8.** Top of levers with release mechanisms **9.** Brass plate denoting signal controlled by lever **10.** Framed diagram of signalling system at Radstock North **11.** Bell tapper to send coded messages to the next signal box **12.** Signal levers **13.** Single line electric key token instrument **14.** Hoops used to pass tokens to drivers **15.** GWR clock **16.** Lamp allows signal box operation at night **17.** Wood-fired stove

EXTERIOR

The signal box is built right next to the level crossing and the railway, with the track connected to the box by complex interlocking mechanics. Each lever inside the box connects to a series of metal pulleys, chains, pivots, and rods that either change a signal, switch a point, or open and close a gate.

How Steam Locomotives Work

The power of steam has long been recognized as a potential energy source and as early as the 1st century CE steam-powered devices appeared in the writings of Hero of Alexandria. It wasn't until the dawn of the industrial era, however, that effective ways were found to harness steam power. In 1712 English ironmonger and inventor Thomas Newcomen developed a steam-powered pump to clear water from mines. Stationary engines such as Newcomen's became mobile when another English engineer,

Richard Trevithick, began experimenting with high-pressure steam engines. These could be made small enough to be mounted on wheels and, for the first time, steam could be used for propulsion. Trevithick's first engine ran on roads, but in 1803 he built a steam locomotive for the Pen-y-Darren colliery that ran on iron track. Within 30 years the railway revolution had begun, providing transport for the masses, and steam was to power the world's railways for more than a century.

CREATING STEAM POWER

To generate steam, hot gases pass from the firebox furnace along tubes that run through the boiler, where they are surrounded by water. The hot "fire tubes" boil the water and steam collects at the top of the boiler. This is referred to as "saturated steam" and a regulator valve controls the rate at which it is fed into the main steam pipe. Superheater pipes then typically boost the steam temperature to give it even more energy before feeding it to the cylinders, where it expands to drive the pistons. Exhaust steam is released through the blast pipe to the chimney, helping to draw hot gases along the fire tubes.

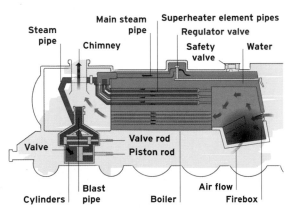

KEY
- Steam exhaust
- Saturated steam
- Superheated steam
- Hot gases

Stoking the firebox
The fireman feeds coal into the firebox when the engine is running, but the fire will have been lit many hours before to raise the temperature slowly and avoid damaging the boiler.

STEAM LOCOMOTIVE COMPONENTS

The essential principles of steam power remained the same throughout the steam age, although locomotives grew more sophisticated. Early steam engines had just one fire tube, for example, but Stephenson's *Rocket* had 25 and later locomotives had 150 or more. Depending on its job – shunting, hauling freight, or a passenger express – a locomotive had to deliver its power in different ways using more pistons or more driven wheels, but the basics remained largely the same.

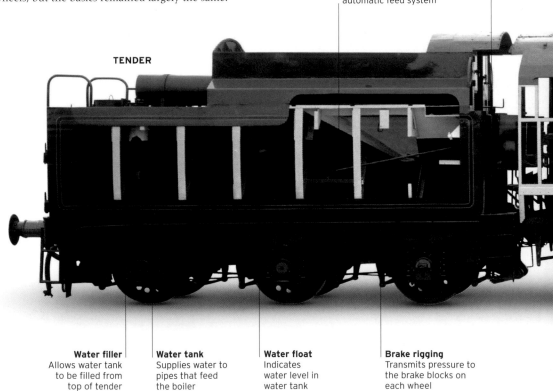

Tender handbrake
Applies the tender brakes when the handle is turned

Coal space
Coal goes from here to the firehole via the fireman's shovel or an automatic feed system

TENDER

Water filler
Allows water tank to be filled from top of tender

Water tank
Supplies water to pipes that feed the boiler

Water float
Indicates water level in water tank

Brake rigging
Transmits pressure to the brake blocks on each wheel

STEAM PROPULSION

Water from the boiler is heated to produce steam, which is then superheated and transferred at high pressure via the steam pipe to the cylinder. Entering the cylinder through a valve, the high-pressure steam pushes a piston which, in turn, drives the series of rods and pivots that turn the driving wheel, thus converting linear motion to rotation.

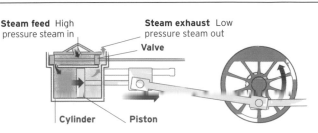

Steam feed High pressure steam in

Steam exhaust Low pressure steam out

Valve

Cylinder   **Piston**

1 Outward stroke High-pressure steam is fed via a valve into the front of the cylinder where it expands and pushes the piston, which rotates the wheel by half a turn

INSIDE THE CAB

Most steam locomotives had a crew of two: fireman and driver. The driver was in charge and controlled the locomotive using the regulator (which acts like a throttle), the reverser, and the brake. Watching his gauges and looking out for trackside signals, the driver regulated the train's speed. The duties of the fireman were to maintain a good supply of steam by stoking the fire, and an adequate level of water by checking the gauge glass. The fireman used the injector control to force water from the tender into the boiler. With the driver, he would also keep an eye out for trackside signals, especially on curves.

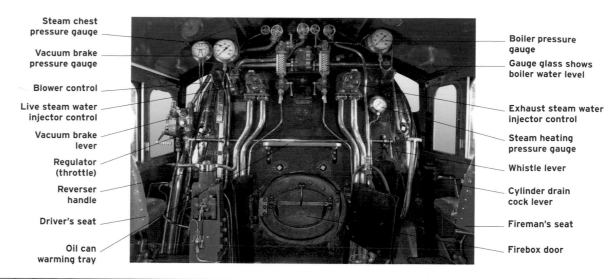

Steam chest pressure gauge
Vacuum brake pressure gauge
Blower control
Live steam water injector control
Vacuum brake lever
Regulator (throttle)
Reverser handle
Driver's seat
Oil can warming tray

Boiler pressure gauge
Gauge glass shows boiler water level
Exhaust steam water injector control
Steam heating pressure gauge
Whistle lever
Cylinder drain cock lever
Fireman's seat
Firebox door

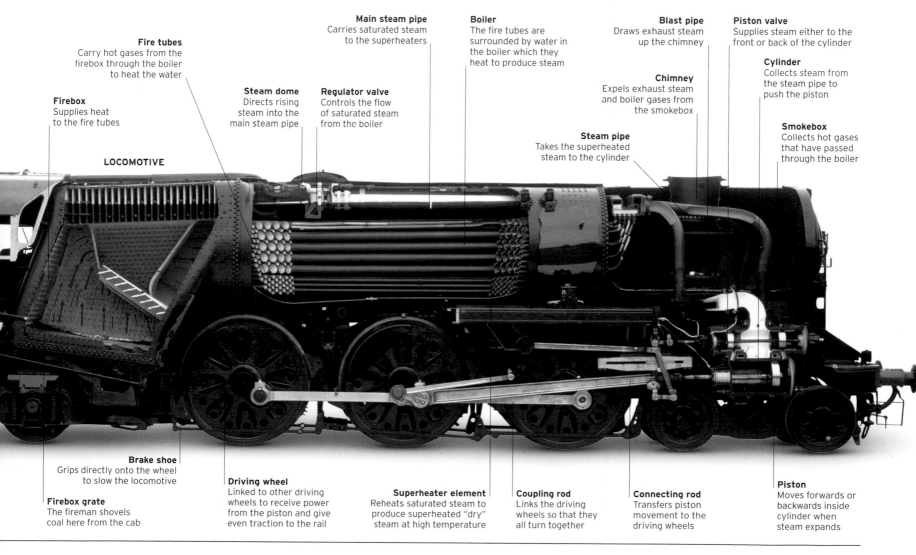

Main steam pipe Carries saturated steam to the superheaters

Boiler The fire tubes are surrounded by water in the boiler which they heat to produce steam

Blast pipe Draws exhaust steam up the chimney

Piston valve Supplies steam either to the front or back of the cylinder

Fire tubes Carry hot gases from the firebox through the boiler to heat the water

Cylinder Collects steam from the steam pipe to push the piston

Firebox Supplies heat to the fire tubes

Steam dome Directs rising steam into the main steam pipe

Regulator valve Controls the flow of saturated steam from the boiler

Chimney Expels exhaust steam and boiler gases from the smokebox

Steam pipe Takes the superheated steam to the cylinder

Smokebox Collects hot gases that have passed through the boiler

LOCOMOTIVE

Brake shoe Grips directly onto the wheel to slow the locomotive

Firebox grate The fireman shovels coal here from the cab

Driving wheel Linked to other driving wheels to receive power from the piston and give even traction to the rail

Superheater element Reheats saturated steam to produce superheated "dry" steam at high temperature

Coupling rod Links the driving wheels so that they all turn together

Connecting rod Transfers piston movement to the driving wheels

Piston Moves forwards or backwards inside cylinder when steam expands

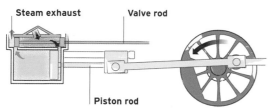

Steam exhaust | Valve rod
Piston rod

2 Exhaust The wheel is connected to the valve via a series of rods. These open the valve to allow the steam, which has now lost pressure, to escape.

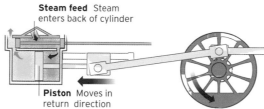

Steam feed Steam enters back of cylinder
Piston Moves in return direction

3 Return stroke The movement of the valve also allows high-pressure steam to enter the back of the cylinder, allowing the return phase of the stroke to begin.

Steam exhaust
Piston Ready for next outward stroke

4 Exhaust Once the wheels have made another half turn, the valve allows spent steam to escape and fresh steam to enter, and the cycle begins again.

How Diesel Locomotives Work

The first diesel engine was demonstrated in 1893 by the German engineer Dr. Rudolf Diesel, who went on to build the first reliable example in 1897. A diesel engine works by drawing air into the cylinders and compressing it to increase its pressure and temperature. Diesel fuel is then injected into it and the resulting combustion produces energy that pushes a piston, which drives a crankshaft. Different transmission systems (electric, mechanical, and hydraulic) are used to transfer the power from the crankshaft to the wheels. A diesel engine can be very powerful; those used in ships can be over 50,000 hp – railway applications are more typically 2,500-4,500 hp. Early diesel locomotives introduced in the 1930s and '40s were cheaper to operate than steam locomotives, especially where oil was plentiful, because they needed far less manpower. Today diesel-powered trains are used worldwide, particularly on less busy lines where electrification is not economical.

DIESEL-ELECTRICS

Most diesel locomotives (and some diesel multiple units) have electric transmissions, and are called "diesel-electric". In a diesel-electric, the power output in the diesel engine uses a transmission system to convert mechanical energy produced by the engine into electrical power. This is achieved by using the engine crankshaft to power a generator (more recently an alternator) to produce electricity. This electrical power in turn operates the traction motors, which are fitted to the wheels or axles of the train. Diesel-electric locomotives are different from electric locomotives – they carry their own power plants rather than relying on an outside supply of electricity.

Diesel-electrics originally ran on DC (direct current) power supplied by a generator, but developments in technology in the 1960s allowed for the use of more reliable AC (alternating current) power supplied by an alternator instead of a DC generator. The AC power from the alternator was passed through a rectifier to transform it to DC electricity to power the traction motors. Advances in traction inverter technology in the 1980s and 1990s allowed the AC supply to power the motors directly, using a system known as three-phase supply.

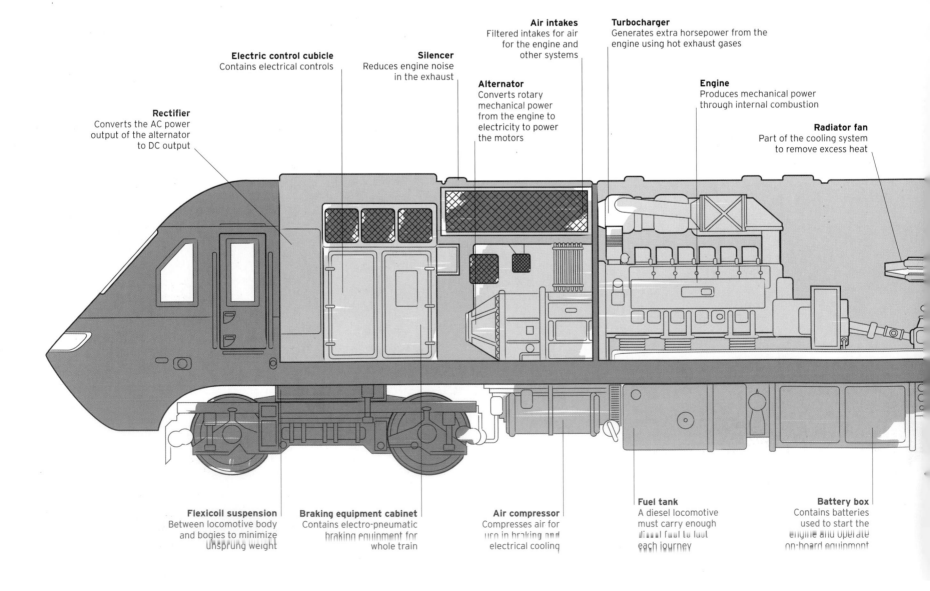

Electric control cubicle
Contains electrical controls

Silencer
Reduces engine noise in the exhaust

Air intakes
Filtered intakes for air for the engine and other systems

Turbocharger
Generates extra horsepower from the engine using hot exhaust gases

Alternator
Converts rotary mechanical power from the engine to electricity to power the motors

Engine
Produces mechanical power through internal combustion

Rectifier
Converts the AC power output of the alternator to DC output

Radiator fan
Part of the cooling system to remove excess heat

Flexicoil suspension
Between locomotive body and bogies to minimize unsprung weight

Braking equipment cabinet
Contains electro-pneumatic braking equipment for whole train

Air compressor
Compresses air for use in braking and electrical cooling

Fuel tank
A diesel locomotive must carry enough diesel fuel to last each journey

Battery box
Contains batteries used to start the engine and operate on-board equipment

DIESEL-MECHANICALS

A mechanical transmission on a diesel locomotive consists of a direct mechanical link between the diesel engine and the wheels. There are two types of mechanism to achieve this. In a direct-drive type mechanism, the engine is connected to the axles via driveshafts, differentials and gearing. The second type is the coupling rod-drive which is used on rigid locomotives that have no pivoting bogies. To maintain efficient adhesion, coupling rods are attached to the outer sides of the wheels of all the powered axles, powering all of the wheels at once.

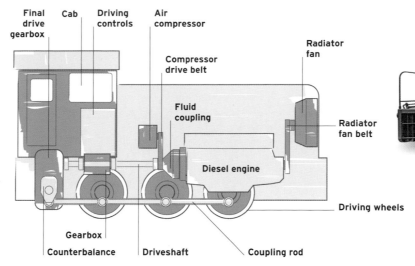

Final drive gearbox · Cab · Driving controls · Air compressor · Compressor drive belt · Radiator fan · Fluid coupling · Radiator fan belt · Diesel engine · Driving wheels · Gearbox · Counterbalance · Driveshaft · Coupling rod

SHUNTER
A shunter or switcher is a small railroad locomotive used for moving trains safely between storage yards and passenger stations. Shunters also assemble freight trains before a hauling locomotive takes over. Many shunters are diesel-mechanical locomotives as they do not need to be capable of high speed.

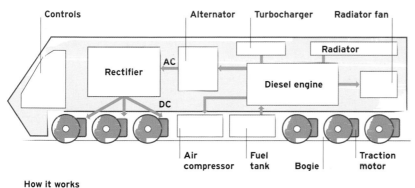

Controls · Alternator · Turbocharger · Radiator fan · Radiator · Rectifier · AC · Diesel engine · DC · Air compressor · Fuel tank · Bogie · Traction motor

How it works
The diagram above shows how power is transferred from the diesel engine to the traction motors on the wheels, through the alternator and rectifier.

DIESEL-HYDRAULICS

Diesel-hydraulic locomotives have similarities to their diesel-mechanical cousins, but while most diesel-mechanical locomotives or diesel-mechanical multiple units are only capable of relatively slow speeds using low-powered engines, diesel-hydraulics are able to operate at higher speeds with much more powerful engines. This is because they have a torque converter instead of a gearbox. The torque converter contains a thick, viscous fluid inside a rotary impeller system to transfer power based on the amount of speed and power the engine is producing. German designers favoured diesel-hydraulics after World War II and large numbers were built; locomotives were even built for export as far afield as the USA and Asia.

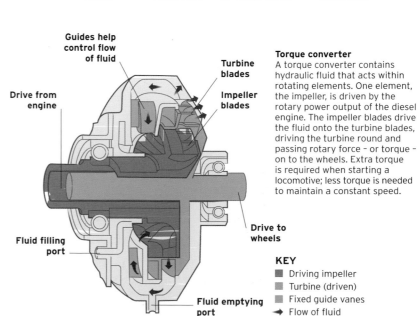

Exhaust · Main air reservoir tanks · Diesel engine · Fuel tank · Cooling coil for air compressors · Batteries

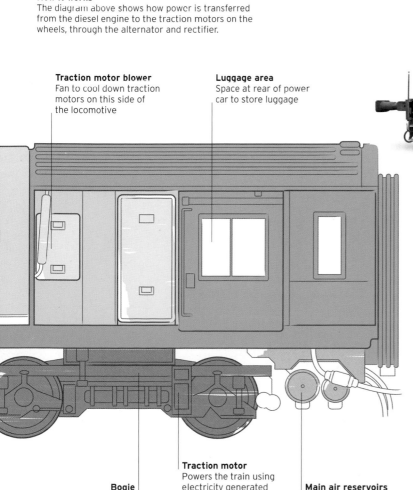

Traction motor blower
Fan to cool down traction motors on this side of the locomotive

Luggage area
Space at rear of power car to store luggage

Bogie
Specially designed for high-speed operation

Traction motor
Powers the train using electricity generated by the alternator; one fitted to each axle

Main air reservoirs
Contain air for braking and other uses

Guides help control flow of fluid · Turbine blades · Impeller blades · Drive from engine · Fluid filling port · Drive to wheels · Fluid emptying port

Torque converter
A torque converter contains hydraulic fluid that acts within rotating elements. One element, the impeller, is driven by the rotary power output of the diesel engine. The impeller blades drive the fluid onto the turbine blades, driving the turbine round and passing rotary force – or torque – on to the wheels. Extra torque is required when starting a locomotive; less torque is needed to maintain a constant speed.

KEY
■ Driving impeller
■ Turbine (driven)
■ Fixed guide vanes
→ Flow of fluid

How Electric Locomotives Work

In Europe, electric trains were initially developed as a more efficient alternative to steam and early diesel locomotives. The first electric locomotive ran in 1879 in Berlin, Germany. However, much of the impetus for the switch to electric traction was driven by the increasing use of railway tunnels, especially in urban areas. In 1890 the first working underground system opened in London using electric locomotives, and electricity soon became the power supply of choice for subways, helped greatly by the introduction of multiple-unit train control in 1897. In the US, electrification of a main line was first used on a 4-mile (6.4-km) stretch of the Baltimore Belt Line of the Baltimore & Ohio Railroad, although electrification was confined to urban areas with dense traffic. The introduction of alternating current as a power supply enabled longer and heavier trains to be operated by electric locomotives and also increased their speed and efficiency.

ELECTRIC TRAINS

Like diesel-electric locomotives, electric trains employ electric motors to drive the wheels but, unlike diesel-electrics, electricity is generated externally at a power station. The current is picked up either from catenaries (overhead cables) via a pantograph, or from a third rail. As they do not carry their own power-generating equipment, electric locomotives have a better power-to-weight ratio and greater acceleration than their diesel-electric equivalents. This makes electric trains ideal for urban routes with multiple stops. They are also faster and quieter than diesel-powered trains. The world rail speed record is held by an electric train – a specially converted French TGV which achieved 357¼ mph (574.8 km/h) in 2007.

Electric locomotive components
For an electric locomotive that is powered via catenary, the pantograph picks up the power supply and transfers it to a transformer, where it is converted to the correct voltage to power the traction motors attached to each wheel. This power allows the locomotive to move.

Air-conditioning unit
Provides air conditioning for driver's cab and electrical equipment

Air reservoirs
Supplies air for traction motor blowers and other compressed-air-cooled electrical equipment

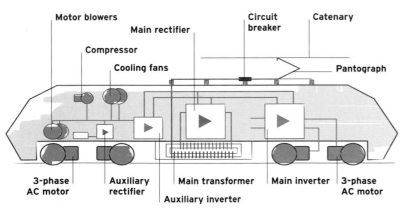

Motor blowers · Main rectifier · Compressor · Cooling fans · Circuit breaker · Catenary · Pantograph · 3-phase AC motor · Auxiliary rectifier · Main transformer · Auxiliary inverter · Main inverter · 3-phase AC motor

How it works
In the three-phase AC electric locomotive above, the AC power supply is converted to lower-voltage DC power by the transformer and rectifiers. Inverters then convert this power back to AC – but at the same lower voltage – to supply power to the motors.

KEY
- High-voltage AC current from catenary
- Converted lower-voltage DC current
- Converted lower-voltage AC current

Smoothing choke
Smooths the DC electric supply to ensure consistency of supply to the motors

Air compressors
Feed the traction motor blowers that help keep the engine cool

THIRD RAIL

Many subway and light rail systems use a third power rail as a method of power supply because it is cheaper to install than overhead lines and is relatively efficient. A shoe extending from the train makes contact with the power rail, and conducts electricity to the train. The system has the advantage that many trains can use it at the same time, disengaging when they no longer need power. The power rail carries a high current that is potentially fatal to humans and animals that come in contact with it, so measures are taken to minimize the risk of contact, especially in stations and depots.

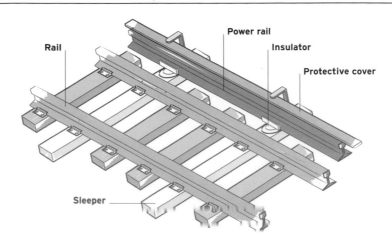

Power rail · Insulator · Protective cover · Rail · Sleeper

Third-rail layout
The power rail lies on insulators mounted on sleepers, and sits alongside the running rails used by the wheels of the train

OVERHEAD LINES

Electric trains that collect their current from catenaries (overhead cables) use a power-collector device such as a pantograph, bow collector, or trolley pole. The power collector is in contact with the lowest overhead wire – the contact wire. Normally made from copper or aluminium, the contact wire is designed to carry several thousand amps of current while remaining in line with the track and withstanding hostile weather conditions. The mechanics of power-supply wiring is not as simple as it looks. The contact wire's tension has to be kept constant; to negotiate curves in the route, for example, the wire has to be held in tension horizontally while it is pulled laterally. The overhead wire is deliberately mounted in a zig-zag pattern to avoid wearing holes in the pantograph.

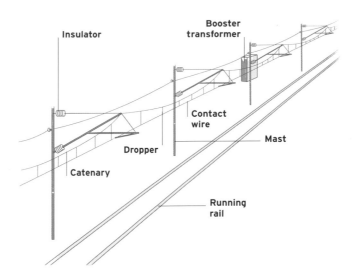

Pantograph
The pantograph is kept in contact with the overhead line using a spring or an air-pressure device. Its contact strips are designed so that they do not get hooked up over the top of the contact wire as the train moves along.

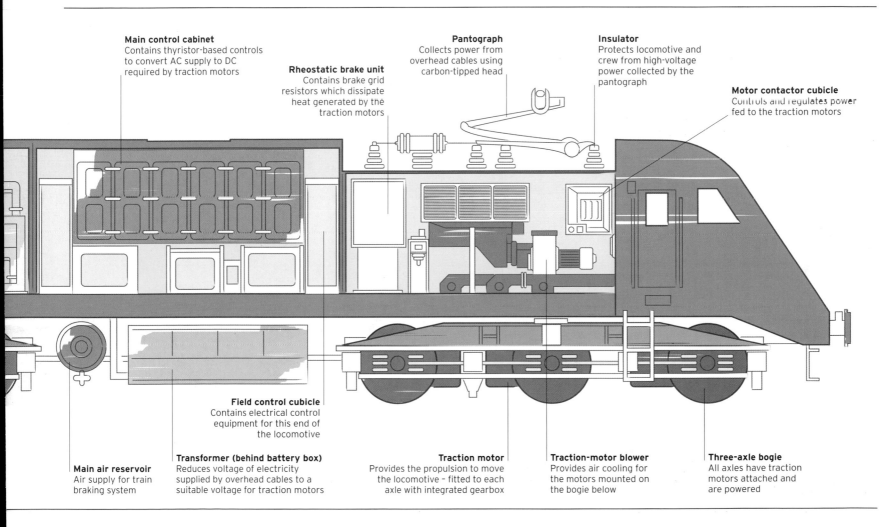

Main control cabinet
Contains thyristor-based controls to convert AC supply to DC required by traction motors

Rheostatic brake unit
Contains brake grid resistors which dissipate heat generated by the traction motors

Pantograph
Collects power from overhead cables using carbon-tipped head

Insulator
Protects locomotive and crew from high-voltage power collected by the pantograph

Motor contactor cubicle
Controls and regulates power fed to the traction motors

Field control cubicle
Contains electrical control equipment for this end of the locomotive

Main air reservoir
Air supply for train braking system

Transformer (behind battery box)
Reduces voltage of electricity supplied by overhead cables to a suitable voltage for traction motors

Traction motor
Provides the propulsion to move the locomotive – fitted to each axle with integrated gearbox

Traction-motor blower
Provides air cooling for the motors mounted on the bogie below

Three-axle bogie
All axles have traction motors attached and are powered

Shoe contact
Trains are fitted with a "shoe" that collects current from the power rail. The simplest design is known as the "top contact", with the pick-up shoe sliding along the top part of the power rail. However, the smallest amount of snow or ice on the exposed rail can render it ineffective. Side contact offers more protection from the elements, but bottom contact is superior because it makes contact with most of the rail and is unaffected by bad weather.

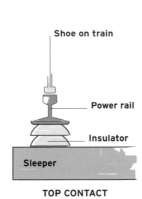

TOP CONTACT

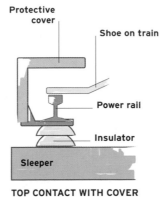

TOP CONTACT WITH COVER

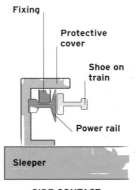

SIDE CONTACT

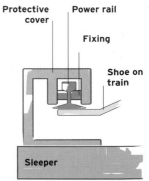

BOTTOM CONTACT

Glossary

Adhesion
The frictional grip between the wheels of a locomotive and the rail of a track, which is affected by axle weight. Particularly important when a locomotive is starting.

Air brake
A braking system that uses compressed air as its operating medium. To apply the brake, the compressed air is released into a cylinder, pushing a piston and spring that push the brake block against the wheel.

Air cushion
A "spring" of air used in modern suspension systems.

Alternating current (AC)
An electric current that reverses its direction of flow rapidly at regular intervals. The rate at which it reverses per second is the frequency, and is calculated in cycles, or Hertz (Hz). See also **Direct current (DC)**.

Alternator
An electromechanical device that converts mechanical energy into electrical energy in the form of alternating current (AC). Used in diesel-electric and electric locomotives.

Articulated locomotive
A locomotive (often steam) with two or more engine units mounted on the same frame but pivoted so that they can move independently of each other. This allows them to transition through curves despite a long wheelbase.

Articulated train
An interconnected train set with cars that are each linked together by a single, pivoting bogie.

Ashpan
Located beneath the firebox of a coal-powered steam locomotive, this pan collects the ash and cinders that fall through the grate of the firebox.

Atlantic
A steam locomotive with a wheel arrangement of 4-4-2 – four leading wheels on two axles, four powered and coupled wheels, and two trailing wheels. First seen in 1880, it was also called a Milwaukee, after the Milwaukee Road, which used the type for its high-speed passenger operations.

Axlebox
A metal casing housing the bearing in which the end of an axle rotates.

Axle load
The fraction of a vehicle's weight that is carried by a given axle. Tracks are designed to carry a maximum axle load.

Ballast
The bed of stone, gravel, or cinders on which a rail track is laid. Sleepers are bedded into the ballast to support the rails. See also **Blanket**, **Formation**, **Subgrade**.

Banker
An extra locomotive that is coupled to a train to help it climb a steep section of track. Known in the US as a helper.

Bar-frame locomotive
A lighter weight steam locomotive originally developed by Edward Bury in 1838, which had a frame made of bars rather than plates. This type was adopted as standard in the US.

Bell code
A language using bell signals to describe trains used by signallers to receive and pass on trains.

Bell tapper
A device used to tap out bell signals between signallers.

Big end
The larger crankpin end of a connecting rod, bigger than the crosshead end as the stresses are greater.

Blanket
An optional layer in the formation of track, the blanket is made of coarse material, and supports the layer of ballast. See also **Ballast**, **Formation**, **Subgrade**

Blastpipe
A pipe that conveys exhaust steam from the cylinders up the chimney of a steam locomotive. This creates a partial vacuum, increasing the flow of air passing through the firebox.

Block
In signalling terms, a section of track that sits between two signals. Trains cannot enter the block if the first signal is "stop".

Bo-Bo
A common axle configuration that describes a locomotive that has two groups of twin-set powered axles. See also **Co-Co**, **Wheel arrangement**

Bogie
A set of pivoted wheels attached to suspension components placed at the front or rear of a locomotive to give guidance and added support. Known in the US as a truck.

Boiler
The part of a steam engine in which steam is produced and circulates. The boiler must be filled with water almost to the top. The water is generally heated by fire tubes, producing steam, which builds to a high pressure. The fireman ensures the boiler is sufficiently filled with water.

Boilerman see **Fireman**

Boxcar see **Van**

Brake
A locomotive has a set of brakes to slow it down, and is normally fitted with an additional control that engages brakes along the length of the train via the brake rigging. Brakes are activated by air, steam, or a vacuum. See also **Air brake**, **Vacuum brake**

Brake block
The friction material that is pressed against a wheel to slow a train down when the brake is applied.

Brake rigging
The system of rods and levers that connect the brake controls to the brake blocks on each wheel.

Brake van
A railcar at the back of a train that provides braking power for goods trains and accommodation for the train guard. Known in the US as a caboose.

Branch line
A secondary railway line that branches off a main line, serving local stations.

Broad gauge
Any gauge in which the rails are spaced more widely than the standard gauge of 4ft 8½ in (1.435 m); for example, Isambard Kingdom Brunel's 7-ft ¼-in (2.14-m) gauge.

Buffer
A device that cushions the impact of rail vehicles against each other.

Buffer stop
The structure at the end of a track that stops a train from travelling any further. Known in the US as a bumper post.

Bullhead rail
A type of rail developed in the UK, in which the top half of the rail mirrors the bottom half. This design was intended to make rails last longer. Once the running side is worn out the rail can be turned over and reused.

Bumper post see **Buffer stop**

Bunker
An enclosure used to store coal at the back of locomotive not followed by a tender.

Bus connector
On an electric multiple unit train, the equipment that transfers the electricity supplied by the catenary from one unit to the next.

Cab
The control room of a locomotive, housing the engine crew.

Cabin car
A railway car used by railway workers to monitor track conditions. It is usually attached to the end of a train.

Caboose see **Brake van**

Cant
The angle of elevation of a rail, relative to vertical or to its partner rail. Known in the US as superelevation.

Car, carriage, coach
Various terms that describe a passenger-carrying rail vehicle.

Catenary
Originally referring to the wire that supported the conductor wire of an overhead electrification system, the term catenary now applies to the entire overhead wire arrangement. Also known as overhead lines and overhead wires.

Chimney
The opening in the top of the smokebox through which exhaust gases and steam escape. Known in the US as a smokestack.

Class
A group of locomotives built to a common design. Can also refer to the level of passenger comfort and service provided on a particular train or carriage, e.g. first class.

Class 1 railroad
A US mainline railway that has annual carrier operating revenues of more than $250 million.

Classification light see **Marker light**

Classification yard see **Marshalling yard**

Co-Co
Refers to any diesel or electric locomotive that has two triple-sets of powered axles. See also **Bo-Bo**, **Wheel arrangement**

Coal space
The portion of a steam locomives tender that carries coal to fuel the firebox. The rest of the tender carries water for the boiler.

Collector shoe
A power collection device attached to an electric train that picks up electricity from an electrified third rail that runs alongside the running track.

Compound locomotive
A steam locomotive that uses two sets of cylinders, the second powered by exhaust steam from the first.

Compression ignition
The process of using heat from compression to ignite and burn fuel in an internal combustion engine. Compression ignition engines are known as diesel engines, and differ from spark ignition engines that use a spark plug to ignite fuel. See also **Diesel**

Conductor see **Guard**

Conjugated valve gear
The operation of a valve on a steam locomotive cylinder by means of levers driven by the motion of the valve gear on two other cylinders. Used by Sir Herbert Nigel Gresley on the three-cylinder locomotives he designed for the Great Northern and the LNER in the UK.

Connecting rod
On a steam engine, a connecting rod links the piston rods to the crankpins of the driving wheels. In some early electric locomotives, the connecting rods linked the crankshaft with the driving wheels.

Consolidation
A locomotive with a 2-8-0 wheel arrangement. It has two leading wheels on one axle, followed by eight powered and coupled driving wheels on four axles. Introduced in the 1860s, it was popular in the US and Europe as a freight hauler.

Container
A metal freight box that can be packed with goods, sealed, and then transported by specially adapted trains, trucks, and ships.

Coupler, Coupling
The mechanism for connecting rail vehicles together. Methods are standardized across a single railway to allow any rolling stock to be coupled together. Known in the UK as a coupling and in the US as a coupler.

Coupling rods
The driving wheels along each side of a steam locomotive are linked together by coupling rods, also known as side rods. Coupling the wheels spreads the power and reduces the possibility of wheels slipping.

Cowcatcher *see* **Pilot**

Crank
The part of a steam locomotive that transmits power from the piston to the driving wheels via connecting rods.

Crankpin
A large steel pin that is pressed into the wheel centre. On steam engines, the driving wheels are driven by rods that transmit force to the wheels through the crankpins.

Crankshaft
In steam locomotives, a shaft that acts upon cranks to convert the linear motion of the piston into rotary motion. This rotary motion drives the wheels.

Crosshead
The point of connection between the piston and the connecting rod that, along with the slidebars, keeps the piston rod in line as it moves in and out of the cylinder.

Cutting
A channel dug through a hillside to enable a rail track to maintain a shallow grade.

Cylinder
An enclosed chamber in which a piston moves to produce power that is transmitted to the wheels. On a steam locomotive, the piston is made to move by the force of high-pressure steam acting against it.

Diesel
Unlike petrol engines, diesel engines use compressed air, rather than a spark, to ignite the oil that fuels them. On a locomotive, the transmission of power from a diesel engine to the wheels may be by electric, mechanical, or hydraulic means. *See also* **Compression ignition**

Diesel-electric
Any locomotive, multiple unit, or railcar that utilizes the diesel-electric system. In a diesel-electric, mechanical power generated by combustion is converted into an electric charge in a generator or alternator, and this electricity powers motors that drive the axles.

Diesel-hydraulic
Any locomotive, multiple unit, or railcar that utilizes the diesel-hydraulic system. In a diesel-hydraulic, power generated by combustion is passed through a torque converter that transfers power to the wheels based on the amount of speed and power the engine is producing.

Diesel-mechanical
Any locomotive, multiple unit, or railcar that utilizes the diesel-mechanical system. In a diesel-mechanical, power generated by combustion is transferred directly to the wheels by means of driveshafts, gearing, and differentials.

Direct current (DC)
An electric current that flows in a constant direction. Alternating current (AC) has significant advantages over direct current in terms of transforming and transmission.

Double-heading
The use of two locomotives, with separate crews, at the head of a train.

Driving wheels
The powered or driven wheels of a locomotive that provide traction.

Dynamic breaking
In electric and diesel-electric locomotives and multiple units, the electric traction motors can be used as generators that act as brakes to slow down the train. Excess energy may be dissipated as heat through brake grid resistors (this is known as rheostatic braking). On an electric train, the excess energy may also be absorbed back into the power supply system (this is known as regenerative braking).

Dynamometer
A device (also called a dyno) used for measuring force, torque, or power. On the railways, dynamometer cars are used to measure a locomotive's speed.

Ejector
Part of a vacuum brake system. The ejector evacuates the brake pipe to create a vacuum, which releases the brakes.

Electrics
Refers to all locomotives, multiple-unit trains, and railcars that draw the electric power for traction from an external source. The electric supply is either picked up from a conductor rail placed beside the track, or from a catenary.

Elevated railway
A railway built on raised platforms. Examples are the former Liverpool Overhead Railway in the UK and part of the New York Subway in the US.

Embankment
A raised pathway across a depression in the landscape that enables a rail track to maintain a shallow gradient.

Engine
The power source of a locomotive, driven by steam, electricity, or diesel. Steam locomotives may also be referred to as steam engines.

Exhaust
The used steam and combusted gases produced by either a steam or a diesel locomotive.

Express train
A train that stops only at certain larger stations on its route in order to arrive at its final destination faster.

Firebox
The section at the rear of a steam locomotive boiler that houses the fire that heats the water in the boiler. Fuel is fed into the firebox from the cab, and the generated heat is fed through the boiler by the fire tubes.

Firehole
The aperture in the firebox of a steam locomotive through which coal or other fuel is fed by the fireman.

Fireman
A crew member responsible for keeping the firebox of an engine fed with coal or other fuel. Also known as a stoker or boilerman.

Fire tubes
Tubes running between a steam locomotive's firebox and smokebox. Hot gases drawn through the fire tubes heat the water surrounding the tubes.

Flange
The projecting lip on the inside edge of a wheel that guides the wheel along a rail.

Flat-bottomed rail
The standard rail used today, which takes the form of a T-shape with a wide, flat base.

Footplate
The floor of a locomotive driving cab where the crew stands. Footplate can also refer to the entire cab.

Formation
The substructure of a track on which the sleepers and rails are laid. *See also* **Ballast**, **Blanket**, **Subgrade**

Freight
A term used to describe trains transporting finished goods and raw materials. It can also refer to the load of materials or products that are being carried.

Funicular railway
Used on tram, cliff, and industrial lines, funicular railways use cables or chains to move vehicles up and down slopes.

Gangway
A flexible structure provided at the ends of coaches to provide access from one coach to another.

Garratt locomotive
An articulated steam locomotive with a boiler in a central frame and two engines on separate frames at each end.

Gas turbine
A type of internal combustion engine that uses high-temperature, high pressure gas to generate energy. Both US and Russian railways are now experimenting with gas turbine-electric locomotives (GTELs), which use a gas turbine to drive an electric generator or alternator.

Gauge
The distance between the inside running edges of the rails of a track. Many gauges are used in different countries and on different railways. Also denotes a visual display of readings for steam, pressure, etc.

Gauge glass
A vertical glass tube in a steam locomotive cab that indicates the water level in the boiler and firebox.

Generator
An electromechanical device that converts mechanical energy to electrical energy in the form of direct current (DC).

Gondola *see* **Open wagon**

Grade, Gradient
The slope of a track. Known in the UK as gradient and in the US as grade.

Grade crossing *see* **Level crossing**

Grate
A grille of firebars at the base of a firebox upon which the fire rests. The gaps in the grille allow in air to assist the fire.

Guard
A member of a train's crew who performs ticketing duties. The guard looks after parcels and other freight in the guard's van, and may also be responsible for the brakes. Known in the US as a conductor, a term which is increasingly used in the UK.

Handcar *see* **Pump trolley**

Helper *see* **Banker**

Horsepower (hp)
A unit of power equal to 550 foot-pounds per second (745.7 watts). Used to express the power produced by steam, diesel, or electric locomotives.

Hot box
Term for an axlebox that has overheated due to inadequate lubrication or too heavy a load.

Injector
A device that feeds water into the boiler of a steam locomotive against the pressure of steam in that boiler.

Interchange
A railway station where passengers can transfer from one train to another that follows a different route. Known in the US as a transfer.

Interlocking tower *see* **Signal box**

Intermodal container
A term used to describe a freight container that can be transferred from one mode of transport to another, such as from a train to a lorry or a ship.

Inverter
A piece of electrical equipment on a diesel-electric or electric locomotive that converts direct current (DC) power supply into an alternating current (AC) supply.

Jacobs bogie
Designed by German railway engineer Wilhelm Jakobs, this is a type of bogie used on articulated railcars and tram vehicles. The bogie is placed between two car body sections, rather than underneath, so that the weight of each car is spread on one half of the bogie.

Journal box
The housing in which the end of an axle turns on a bearing.

Kriegslok
Short for *Kriegslokomotive*, this is a German war locomotive. Built in large numbers during World War II, they were cheap and easy to build, easy to maintain, and could withstand extreme weather conditions.

Leading wheel
A wheel located in front of the driving wheels of a steam locomotive that provides support but which is unpowered.

Level crossing
A location where a railway crosses a road or path at the same elevation. Known in the US as a grade crossing or a railroad crossing.

Level junction
A railway junction where multiple lines intersect, crossing the path of oncoming rail traffic at the same elevation.

Light rail
A form of rail transport typically operating within urban environments. Light rail vehicles (LRVs) include streetcars and trams.

Link valve gear
A design of valve gear, designed at the Stephensons' locomotive works in 1842.

Livery
Distinctive colours, insignia, and other cosmetic design features of a rail vehicle.

Loading gauge
The dimensions that a rail vehicle must not exceed, to avoid collisions with trackside objects and structures. Different countries have different loading gauges.

Locomotive
A wheeled vehicle used for pulling trains. Steam and diesel locomotives generate their own power, while electric locomotives collect electricity from an external source.

Maglev train
A train that works by being levitated above and propelled over special tracks by electromagnetic force. Maglevs produce virtually no friction, and are very quiet in operation at high speed.

Main line
An important railway line, often running between major towns or cities.

Marker light
Particularly in the US, a light that was used to signal the status of the train to other drivers. Green marker lights indicated a regularly scheduled train; white marker lights indicated an extra train; and red marker lights attached to the final car indicated the end of the train. Red lights are still used in tail lights around the world today. Also known as a classification light.

Marshalling yard
A place where freight trains are assembled, or where freight wagons for different destinations are moved to the correct train. Known in the US as a classification yard.

Metre gauge
A railway track with the inside of its rails 3 ft 3 in (1 m) apart.

Metro
Internationally, a name that is popularly used for an underground rapid transit system – a type of high-capacity rail public transport in urban areas. Generally known as a subway in the US. Each system has its own name, such as London Underground, New York Subway, and Paris Métro.

Monorail
A railway system based on a single rail. A monorail is often elevated above the ground, and built in urban areas.

Motion
In railway terminology, the collective term for the piston rods, connecting rods, and valve gear of a locomotive.

Motive power depot *see* **Running shed**

Multitube boiler
A locomotive boiler with multiple tubes, which revolutionized steam locomotive design. Stephenson's *Rocket* was the first engine to have a multitube boiler – with 25 copper tubes instead of a single flue or twin flue.

Multiple unit (MU)
A term used in diesel and electric traction that refers to the semi-permanent coupling of several powered and unpowered vehicles to form a single train.

Narrow gauge
Any railway with a gauge narrower than the standard 4 ft 8½ in (1.435 m).

Oil firing
A method of firing a steam locomotive using oil as fuel.

Open wagon
An open-top piece of rolling stock used to transport loose materials such as ore and coal. Known in the US as a gondola.

Overhead lines *or* **Overhead wires** *see* **Catenary**

Pacific
A locomotive with a wheel arrangement of 4-6-2. It has four leading wheels on two axles, six powered and coupled driving wheels on three axles, and two trailing wheels on one axle. The Pacific was a common type of steam passenger locomotive during the first half of the 20th century.

Pantograph
An assembly on the roof of an electric locomotive or electric multiple-unit power car that draws current from an overhead wire (catenary). Also known as a current collector.

Passenger train
A train with carriages intended to transport people rather than goods. These trains travel between stations at which passengers may embark or disembark.

Passing loop, passing siding
A position on a single-track railway where trains travelling in opposite directions can pass each other. Known as a passing loop in the UK and as a passing siding in the US.

Permanent way
The rails, sleepers, and subgrade of a railway line. The term comes from the fact that temporary lines were laid during railway construction, which were then replaced by a "permanent way".

Pilot
A sloping plate or grid fitted to the front of a locomotive; it is designed to push obstructions off the track. Known in the US as a cowcatcher.

Piston
The cylindrical assembly that moves back-and-forth inside each cylinder of a steam or diesel engine. The movement of the piston provides mechanical power, which is transferred by various means to the wheels.

Piston rod
The rod linking the piston in a cylinder with the crosshead.

Points
A track mechanism at the point where two tracks diverge that allows a train to move from one track to another. Known in the US as a switch.

Pullman car
A luxury railway carriage. Pullmans were initially introduced in the US by George Pullman in 1865 as sleeping cars on long-distance trains.

Pump trolley
A small, open railway vehicle propelled by its passengers, often by means of a hand pump. Known in the US as a handcar.

Rack railway
A railway with an additional toothed rack-rail. A train or locomotive running on the railway is fitted with a cog that lines with the teeth on the rail, enabling it to climb slopes that would be impossible for a normal train.

Railcar, railmotor
A self-propelled passenger vehicle, usually with the engine located under the floor.

Railway standard time
Before the introduction of railway timetables, different places in the same country often had their own local time. In the 1840s, railways began to introduce a standardized railway time to avoid confusion caused by local time differences.

Rectifier
A piece of electrical equipment on a diesel-electric or electric locomotive that converts an AC power supply into a DC power supply. They are also used alongside railways to convert traction current.

Regenerative brake *see* **Dynamic braking**

Regulator
A lever used by the driver of a steam locomotive to control the supply of steam to the cylinders. Known in the US as a throttle.

Reverser
Mechanism with a wheel or lever that controls the forward and reverse motion of a steam locomotive.

Rheostatic brake *see* **Dynamic braking**

Rolling stock
A term used by railway companies to refer to the collection of vehicles that run on their railway.

ROD
Stands for the Railway Operating Division of the British Royal Engineers, who maintained the railways in theatres of war during World War I.

Running board, running plate
The footway around a locomotive's engine compartment or boiler.

Running gear
The parts involved in the movement of an engine. Includes wheels, axles, axleboxes, bearings, and springs.

Running shed
An old name for a motive power depot, where locomotives are stored, repaired, and maintained when not in use.

Saddle tank
A tank locomotive that has the water tank mounted on top of the boiler.

Safety valves
In a steam locomotive boiler, relief valves that are set to lift automatically to allow steam to escape if the boiler pressure exceeds a set limit.

Saloon
A luxurious railway carriage used as a lounge, or with private accommodation.

Sandbox *see* **Sanding**

Sanding
The application of sand between the wheel tyres and the rails to increase grip and prevent wheelslip. The sand is piped from a sandbox, which is often situated on top of the boiler.

Saturated steam
Steam that has yet to be superheated to remove any remaining water droplets. Also known as "wet steam".

Semaphore signalling
A system that relies on pivoting arms to relay a signal to drivers. The angle of each pivoting arm tells the driver whether the signal is "stop", "caution", or "all clear".

Shoe *see* **Collector shoe**

Shunter
A small locomotive used for moving trucks or wagons around in a marshalling yard. Known in the US as a switcher.

Shuttle
A railway service that operates between two stations, often without intermediate stops. A common use of shuttle services is to take passengers between airport termini, or from an airport to a city centre.

Side rods *see* **Coupling rods**

Siding
A section of track off the main line used for storing rolling stock.

Signal
A mechanical or electronic fixed unit with an arm or a light that indicates whether a train should stop, go, or use caution.

Signalling token
A token used in old signalling systems. The token was collected by the train's crew at the beginning of a block of track. The token was returned to the signaller at the other end of the "block" of track. This system ensured that at any time, only one train would be travelling within a block.

Signaller, signalman
In the UK, a person employed by a railway to manage and operate the points and signals on a section of track from a signal box. Known in the US as a towerman; in the US the term signaller denotes a signal maintenance worker.

Signal box
A control room in which the movement of trains is controlled by means of signals and blocks, ensuring trains travel safely and to schedule. Known in the US as a tower or interlocking tower.

Sleeper
The cross-piece supporting the rails, made out of wood, concrete, or steel. Early railways also used stone sleeper blocks

Known in the US as a tie or crosstie. The term "sleeper" can also describe a coach or train that provides beds for passengers on overnight or long-distance journeys.

Sleeping car
A carriage with beds where passengers can sleep while travelling. Sleeping cars were first introduced in the US in the 1830s.

Slidebars
On a steam locomotive, slidebars combine with the crosshead to guide the movement of piston rods.

Slip coach
A coach that could be uncoupled from a moving express train and braked to a halt at a station. This allowed passengers to disembark without halting the main train.

Smoke deflectors
Metal sheets attached to the smokebox to funnel air upwards, forcing smoke and steam emitted from the chimney away from the cab to improve visibility.

Smokebox
The leading section of a steam locomotive boiler assembly that houses the main steam pipes to the cylinders, the blastpipe, the chimney, and the ends of the firetubes. Ash drawn through the firetubes collects here.

Smokestack *see* **Chimney**

Spiral
A railway formation in which tracks cross over themselves as they ascend a mountain.

Splasher
A semi-circular guard used to enclose the top section of a large-diameter driving wheel. Often fitted when a wheel protrudes above the running board of a locomotive.

Standard gauge
Rails spaced 4 ft 8½ in (1.435 m) apart. Standard gauge is the most commonly used gauge worldwide. Designed by Robert Stephenson for the first inter-city railways, it is also known as Stephenson's gauge.

Steam chest
The internal part of a locomotive's cylinder block where the valve chamber connects with the steam supply and exhaust pipes.

Steam dome
A chamber on top of the barrel of a steam locomotive's boiler where superheated steam collects and is directed to the cylinders through the steam pipe.

Steam locomotion
Steam locomotion is founded on the principle that when water is heated above its boiling point, it turns to steam and its volume becomes 1,700 times greater. If this expansion takes place within a sealed vessel such as a boiler, the pressure of the steam will become a source of energy.

Steam pipe
The pipe that connects the steam dome to the steam chest in the cylinder block.

Stoker *see* **Fireman**

Streamliner
A locomotive or train set that incorporates streamlining into its shape to provide reduced air resistance.

Subgrade
Ground prepared to give a consistent gradient to tracks that will be laid above it. *See also* **Ballast**, **Blanket**, **Formation**

Subway *see* **Metro**

Supercharging
A way of introducing more air into the cylinders of a diesel engine, by using a turbocharger to force air through the inlet valves at higher than atmospheric pressure.

Superelevation *see* **Cant**

Superheated steam
Steam that has been raised in temperature and volume by adding extra heat as it passes between the boiler and the cylinders. This dries the steam by turning remaining water droplets into gas, thus delivering more power.

Switch *see* **Points**

Switcher *see* **Shunter**

Tail light
The lamp at the rear of a train. In the UK, a train is not complete without a red rear warning light. *See also* **Marker light**

Tank locomotive
A steam engine that carries its fuel and water on its chassis rather than in a tender. The water is often held in side tanks or in saddletanks that encase the boiler.

Telegraph (electric)
A communication system developed in the 1830s that used electrical impulses travelling through wires to send messages. It became the standard instrument of railway communication worldwide.

Tender
A vehicle attached to a steam engine that carries the fuel and water.

Third rail
A system that provides an electric train with power through a conducting third rail set alongside the running tracks. The power is collected via a shoe attached to the train.

Three-phase system
A system that enables a steady supply of AC current without fluctuations to power traction motors, enabling higher traction power to be achieved.

Throttle *see* **Regulator**

Tie *see* **Sleeper**

Tilting train
A train that can lean into bends, enabling it to travel faster around curves without passenger discomfort.

Tower *see* **Signal box**

Towerman *see* **Signaller**

Track
The permanent fixtures of rails, ballast, fastenings, and underlying substrate that provide a runway for the wheels of a train.

Traction
In railway terms, a force that relies on friction between a wheel and a rail to generate motion. *See also* **Adhesion**

Traction motor
An electric motor that uses incoming electrical energy to power the axles. Used in both diesel-electric and electric traction.

Tractive effort
A measure of a locomotive's pulling power; the effort that it can exert in moving a train from standstill. This force is calculated by measuring the energy the locomotive exerts on the rails. *See also* **Traction**

Trailer, trailer car
A passenger vehicle in a multiple unit that has no power traction equipment, and which is powered by the vehicles that are attached to it.

Trailing wheel
A wheel located behind the driving wheels of a steam locomotive that provides support but which is unpowered.

Train
Passenger or freight vehicles coupled together and travelling as one unit along a railway line. Trains can be self-propelled or locomotive-hauled.

Transfer *see* **Interchange**

Transmission
In a diesel locomotive, the method by which power is transmitted from an engine to an axle or the wheels. Transmission may be electrical, hydraulic, or mechanical.

Truck
A small rail wagon. Also, the US term for a bogie.

Turntable
A device for rotating rail vehicles so they can travel back in the direction they came from. Largely obsolete today.

Twin-track railway
A railway that runs two tracks along the same line, each track taking trains in opposite directions, rather than both directions being serviced by a single track.

Underground *see* **Metro**

USATC
An abbreviation that stands for United States Army Transportation Corps. Locomotives built in the US for the USATC were shipped to Europe for use by the Allies in World War II.

Vacuum brake
A type of brake that is held off by a partial vacuum and applied when air is let into the system. Vacuum brakes were used in the UK because, unlike air brakes, they did not require a separate pump.

Valve
In a steam locomotive, valves co-ordinate the movement of steam into and out of the cylinders. In a diesel engine, valves control fuel intake and expulsion of exhaust gases.

Valve gear
Linkages that connect the valves of a steam locomotive and control the movement of the valves.

Van
A flat-bottomed freight wagon with sliding doors on each side. Known in the US as a boxcar.

Vertical cylinder
Vertically mounted cylinders used in early locomotives such as the Stephensons's *Locomotion No.1* and, later, in specialized forms of shunting engines and narrow-gauge locomotives.

Wagon
A general term for a rail vehicle that carries freight.

Walschaerts valve gear
A form of link motion valve gear first patented in 1844 by Egide Walschaerts, a Belgian engineer. It was widely used in Europe, being easier to maintain and lighter than Stephenson's link valve gear. It first appeared in the US in 1876 and was also used extensively there.

Water column, water plug
A hollow pole fitted with a hose and connected to a water supply for filling locomotive water tanks. Water columns may be fitted onto cranes with movable arms to allow water to be supplied to locomotives on either of two adjacent tracks. Known in the UK as a water column, and in the US as a water plug.

Westinghouse brake
A widely used automatic air brake invented in the 1870s by US engineer George Westinghouse. Universally adopted in the US, it was also developed worldwide.

Wet steam *see* **Saturated steam**

Wheel arrangement
A method of classifying locomotives by the distribution of different types of wheels. For steam locomotives, Whyte notation is a common system. Diesel and electric locomotives and powered cars are categorized by the number of powered and unpowered axles that they have. The unpowered axles, which often carry the leading and the trailing wheels, are listed numerically, while the powered axles supporting the driving wheels are given an alphabetical description. *See also* **Bo-Bo**, **Co-Co**, *and* **Whyte notation**.

Wheel unit *see* **Bogie**

Wheelset
An assembly that consists of two wheels attached to an axle on a rail vehicle.

Whyte notation
A classification of steam locomotive wheel arrangements that is based on the number of leading, driving, and trailing wheels. For example, a wheel arrangement of 4-4-0 would denote a locomotive with four leading wheels, four driving wheels, and no trailing wheels.

Yard
An area off the main line used for storing, sorting, loading, and unloading vehicles. Many railway yards are located at strategic points along a main line. Large yards may have a tower from which marshalling operations are controlled.